supporting expeditionary aerospace forces

T0146352

Alternatives for Jet Engine Intermediate maintenance

mahyar A. Amouzegar

Lionel A. Galway

Amanda Geller

Prepared for the United States Air Force

Project AIR FORCE

RAND

The research reported here was sponsored by the United States Air Force under Contract F49642-01-C-0003. Further information may be obtained from the Strategic Planning Division, Directorate of Plans, Hq USAF.

Library of Congress Cataloging-in-Publication Data

Amouzegar, Mahyar A.
  Supporting expeditionary aerospace forces : alternatives for jet engine intermediate maintenance / Mahyar A. Amouzegar, Lionel A. Galway, Amanda Geller.
    p. cm.
  "MR-1431."
  Includes bibliographical references.
  ISBN 0-8330-3103-1
  1. United States. Air Force—Facilities. 2. Jet engines—United States—Maintenance and repair. I. Galway, Lionel A., 1950– II. Geller, Amanda. III. Title.

UG634.49 .A46 2002
358.4'183—dc21

                                                    2001048906

RAND is a nonprofit institution that helps improve policy and decisionmaking through research and analysis. RAND® is a registered trademark. RAND's publications do not necessarily reflect the opinions or policies of its research sponsors.

Published 2002 by RAND
1700 Main Street, P.O. Box 2138, Santa Monica, CA 90407-2138
1200 South Hayes Street, Arlington, VA 22202-5050
201 North Craig Street, Suite 102, Pittsburgh, PA 15213-1516
RAND URL: http://www.rand.org/
To order RAND documents or to obtain additional information, contact Distribution Services: Telephone: (310) 451-7002; Fax: (310) 451-6915; Email: order@rand.org

This report documents research undertaken in support of emerging Air Force employment strategies associated with the Expeditionary Aerospace Force (EAF). EAF concepts turn on the premise that rapidly tailorable, quickly deployable, immediately employable, and highly effective air and space force packages can serve as a credible substitute for permanent forward presence. The success of the EAF will depend, to a great extent, on the effectiveness and efficiency of the Agile Combat Support (ACS) system.

This study is one of a series of RAND reports that address ACS issues in implementing the EAF. Other reports in the series include the following:

- *Supporting Expeditionary Aerospace Forces: An Integrated Strategic Agile Combat Support Planning Framework,* Robert S. Tripp et al. (MR-1056-AF). This report describes an integrated ACS planning framework that can be used to evaluate support options on a continuing basis, particularly as technology, force structure, and threats change.

- *Supporting Expeditionary Aerospace Forces: New Agile Combat Support Postures,* Lionel Galway et al. (MR-1075-AF). This report describes how alternative resourcing of forward operating locations (FOLs) can support employment time lines for future EAF operations. It finds that rapid employment for combat requires some prepositioning of resources at FOLs.

- *Supporting Expeditionary Aerospace Forces: An Analysis of F-15 Avionics Options,* Eric Peltz et al. (MR-1174-AF). This report examines alternatives for meeting F-15 avionics maintenance requirements across a range of likely scenarios. The authors evaluate investments for new F-15 avionics intermediate-maintenance shop test equipment against several support options, including deploying maintenance capabilities with units, performing maintenance at forward support locations (FSLs), and performing all maintenance at the home station for deployment units.

- *Supporting Expeditionary Aerospace Forces: A Concept for Evolving the Agile Combat Support/Mobility System of the Future*, Robert S. Tripp et al. (MR-1179-AF). This report describes the vision for the ACS system of the future based on individual commodity study results.

- *Supporting Expeditionary Aerospace Forces: Expanded Analysis of LANTIRN Options*, Amatzia Feinberg et al. (MR-1225-AF). This report examines alternatives for meeting Low-Altitude Navigation and Targeting Infrared for Night (LANTIRN) support requirements for EAF operations. The authors evaluate investments for new LANTIRN test equipment against several support options, including deploying maintenance capabilities with units, performing maintenance at FSLs, and performing all maintenance at continental Untied States (CONUS) support hubs for deploying units.

- *Supporting Expeditionary Aerospace Forces: Lessons from the Air War over Serbia*, Amatzia Feinberg et al. (MR-1263-AF). This report describes how the Air Force's ad hoc implementation of many elements of an expeditionary ACS structure to support the air war over Serbia offered opportunities to assess how well these elements actually supported combat operations and what the results imply for the configuration of the Air Force ACS structure. The findings support the efficacy of the emerging expeditionary ACS structural framework and the associated but still-evolving Air Force support strategies.

This report documents our work on alternative concepts for Jet Engine Intermediate Maintenance (JEIM) to determine whether peacetime and wartime jet engine maintenance is better performed by JEIM shops located with the aircraft or by organizations operating in a centralized facility. The research addressed in this report was conducted within the Resource Management Program of Project AIR FORCE as one element of a project entitled "Implementing an Effective Air

Expeditionary Force." The project was sponsored by the Air Force Deputy Chief of Staff for Installations and Logistics (AF/IL). This report should be of interest to logisticians and operators in the Air Force concerned with implementing the EAF concept.

The cutoff date for this research is September 2000.

PROJECT AIR FORCE

Project AIR FORCE, a division of RAND, is the Air Force federally funded research and development center (FFRDC) for studies and analyses. It provides the Air Force with independent analyses of policy alternatives affecting the development, employment, combat readiness, and support of current and future aerospace forces. Research is performed in four programs: Aerospace Force Development; Manpower, Personnel, and Training; Resource Management; and Strategy and Doctrine.

CONTENTS

FIGURES

TABLES

Since the end of the Cold War, the U.S. Air Force has been required to perform numerous overseas deployments, many on short notice, in support of crises ranging from humanitarian relief to Operation Desert Storm. To meet these challenges, the Air Force has begun to reorganize itself into an Expeditionary Aerospace Force (EAF) that replaces the forward presence of airpower with a force that can deploy quickly from the continental United States (CONUS) to forward operating locations (FOLs) in response to a crisis; commence operations immediately on arrival; and sustain those operations as needed. This radical change in operational concept requires a similarly radical rethinking of all *support* concepts, including munitions, fuel, housing, avionics repair, and jet engine repair.

Traditionally, the Jet Engine Intermediate Maintenance shop (JEIM) has been located at the FOL. The goal of this project was to evaluate several different alternatives to accomplishing expeditionary JEIM support.[1] We examined the performance of the following alternatives in supporting forces deployed to a notional 100-day major theater war (MTW):

---

[1]In fact, the Air Force has had centralized jet engine repair for various engines several times in its history (e.g., in support of Kosovo operations in 1999), albeit with varying success.

- **Decentralized-deployed (DecDep).**[2] JEIM support is decentralized to each base in peacetime; part of each base's JEIM deploys with the aircraft to the FOL in war.

- **Decentralized–no deployment (Home).** JEIM support is decentralized to each base in peacetime; during war, each base JEIM supports its own deployed unit.

- **Decentralized–forward support location (FSL).** JEIM support is decentralized to each base in peacetime; during war, a single JEIM is set up in theater to support all deployed units with a given type of engine.

- **CONUS support location–forward support location (CSL-FSL).** All units are supported in peacetime by a single centralized JEIM in CONUS; personnel from the CSL deploy to a theater FSL to support deployed units in war. In theater, this option is identical to the previous one.

- **CONUS support location (CSL).** All units are supported in peacetime or during war by a single centralized JEIM in CONUS.

Because none of these alternatives has actually been implemented for CONUS-wide engine fleets or used in an MTW, we developed a simulation model of centralized and decentralized JEIM operation and used this model, supplemented by data analysis from Air Force maintenance databases and unit visits and interviews, to evaluate the alternatives listed above.

We evaluated the performance of each of the alternatives for three different engines: the F100-220, the F100-229, and the TF-34. We sized the alternatives so that no required wartime sorties were missed and then compared the spares levels over the 100-day course of our notional MTW as well as the personnel and transportation required.

Figure S.1 illustrates a representative result: the relative performance in maintaining spares levels over a 100-day conflict of the JEIM alternatives for F100-220 repair for the F-15s deployed to the MTW.

---

[2]We label each alternative in terms of "peacetime repair–wartime repair." For example, the decentralized-deployed case implies a decentralized mode of repair during peacetime (and for nonengaged forces) at home units and deployed JEIM shops at the FOLs.

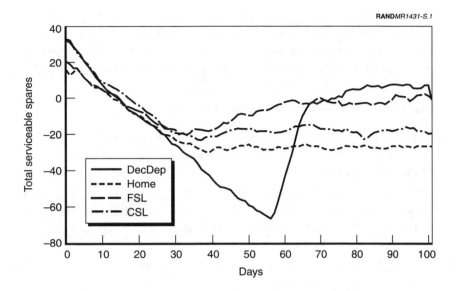

RAND*MR1431-S.1*

Figure S.1—F100-220 F-15 Serviceable Spares

Deploying the JEIM (decentralized-deployed) gives the worst performance in terms of spares: At day 60, these forces are short more than 60 engines, which translates into more than 30 F-15s (out of 60) being not mission capable by virtue of having no operating engines. Because the deployed JEIM takes 60 days to reach complete functionality (owing to its planned deployment schedule and to the time required to assemble the engine test cell), this alternative requires a large number of spares to prevent a buildup of aircraft without engines. In contrast, the FSL option maintains a much higher level of spares in the deployed force with fewer maintenance personnel.[3] These results generally hold for both F100 engines. For the TF-34, which has a much lower removal rate, the difference is much smaller.

However, the centralized alternatives require dedicated and responsive transportation, and their performance is quite sensitive to devia-

---

[3]The CSL-FSL and decentralized-FSL alternatives are represented by the FSL curve because these two alternatives are identical within the theater of operations.

tions from the times assumed in our modeling. For example, we assumed that transportation to and from the FSL is two to four days. Even a small increase causes a loss in combat sorties. However, the FSL alternative is more robust than are the other options to changes in removal rate or in the scenario.

We reached the following major conclusions:

- For support of a fast-breaking MTW, the option of deploying the JEIM to the FOL is too slow to provide adequate support with acceptable spares levels.

- For the F100-220 and F100-229, locating the JEIM at an FSL is the best alternative. However, this requires dependable intratheater transportation. For the TF-34, either an FSL or a CSL provides acceptable performance.

- In peacetime, centralizing repair for small F100-220 bases could provide some resource savings. For the TF-34, a CSL would be effective.

Our analyses show the feasibility of these options, but a number of qualitative issues are also important. These include the support of flight-line engine personnel if the JEIM is not located on base; the issue of control of JEIM resources which are supporting multiple bases; and the question of the transition to wartime if a decentralized structure is retained for peacetime but FSL support is planned for a conflict.

ACKNOWLEDGMENTS

The research documented in this report could not have been done without the active cooperation and help of a large number of Air Force personnel. In these acknowledgments, we list the positions of these personnel as of the time the study was done.

Much of the RAND work on EAF support has been done under the sponsorship of several different Deputy Chiefs of Staff for Installations and Logistics (AF/IL). Our point of contact with IL was primarily Susan O'Neal (AF/ILX), who, together with her staff, provided much useful information and contacts with other Air Force organizations. As described in the report, this particular project was cosponsored by Air Combat Command (ACC/LG) Major General Dennis Haines. We are particularly indebted to ACC/LGSP for their help and involvement as our point of contact with ACC, especially Colonel Stanley Stevens, Lieutenant Colonel John Cooper, Chief Master Sergeant Hank Houtman, Chief Master Sergeant Michael Kinser, and ACC Command Engine Manager Tom Smith. Data were also provided by ACC/LGP under the direction of Ed Merry.

At San Antonio Air Logistics Center, we would like to thank Robert May (SA-ALC/LR) and his staff, particularly Melissa Tinscher and Chris Szczepan, for information on engine requirements. We also benefited from discussions with SA-ALC/LPF; Colonel Patrick Doumit and his staff, especially Colonel Robert McMahon; Greg Hall; Bruce Eberhard; and David Crowley.

For access to and patient help with data from the Comprehensive Engine Management System (CEMS), we are grateful to Charlie Osborn, Phil Garrity, Jim Blain, David Addison, and Walt Cooper. For help gaining access to and interpreting data from the Reliability and Maintainability Information System (REMIS), we appreciate the help of Richard Enz and Thomas Recktenwalt.

We had extensive help from a number of Jet Engine Intermediate Maintenance shops (JEIM) and their senior noncommissioned officers (NCOs). These include Senior Master Sergeant John Kasprak at the 1st Wing (FW) (Langley Air Force Base), Senior Master Sergeant Richard McDyer at the 20th FW (Shaw Air Force Base), Senior Master Sergeant Rod Dottin at the 347th FW (Moody Air Force Base), Senior Master Sergeant Mark Travis at the 366th FW (Mountain Home Air Force Base), Chief Master Sergeant Duane Mackey at the 48th FW (Lakenheath Air Base), Senior Master Sergeant Alex Gasper at the 31st FW (Aviano Air Base), and Chief Master Sergeant Michael Holas at the 52nd FW (Spangdahlem Air Base). All of these people and their staffs graciously organized tours of their shops as well as meetings with their supervisors; collected data and information on their operations; and fielded clarification questions after our departure. This project also utilized a discussion with Chief Master Sergeant Roy Hauck at U.S. Air Forces Europe (USAFE) Headquarters on engine support during Operation Noble Anvil.

Finally, we appreciate the hospitality of the Air Force Research Laboratory at Wright-Patterson Air Force Base, Ohio, for a day of discussions of new propulsion technologies.

As always, we benefited greatly from the comments and constructive criticism of many RAND colleagues, including (in alphabetical order), Frank Camm, Carl Dahlman, Amatzia Feinberg, Louis Miller, Nancy Moore, Richard Moore, Timothy Ramey, C. Robert Roll, Hyman Shulman, and Alan Trevor. We owe special thanks to Robert Tripp, the project leader for all of the EAF support tasks, for his guidance and support. Roger Madison provided programming support to some of the data analyses. In particular, colleagues Louis Miller and Jim Bigelow provided detailed reviews of drafts of this work that substantially improved the presentation.

Our modeling in this project was done with Extend simulation software. We had good technical support from author Bob Diamond and his people at Imagine That! in San Jose, California.

Much of the RAND EAF support work has been done in partnership with the Air Force Logistics Management Agency (AFLMA) at Gunter Air Force Base, Alabama. During this project, Colonel Richard Bereit commanded the agency and helped build the relationship. We especially thank Captain Al Hardemon for his insight on engine maintenance and Master Sergeant Joel Doran for some quick work with load plans for different engines for the U.S. Air Force airlift fleet. As with all of the other EAF work, this project was greatly enhanced by the input of Chief Master Sergeant John Drew. He gave us much good advice, traveled with us on most of our visits, suggested questions, and was in general an invaluable colleague.

It will be evident from our report that there is much discussion in the Air Force engine community on the issues we have studied, and many of the people acknowledged here may have different views and disagree with one or more of our conclusions. Nevertheless, they provided us with full and open access to their facilities, data, and people so that we could pursue our research as we wished. The analysis and conclusions are our responsibility.

ACC	Air Combat Command
ACS	Agile Combat Support
AEF	Aerospace Expeditionary Force
AETC	Air Education and Training Command
AF/IL	Air Force Deputy Chief of Staff for Installations and Logistics
AFLMA	Air Force Logistics Management Agency
AFR	Air Force Reserve
ALC	Air Logistics Center
ANG	Air National Guard
CAMS	Core Automated Maintenance System
CEMS	Comprehensive Engine Management System
CIRF	Centralized Intermediate Repair Facility
CONUS	Continental United States
CSL	CONUS support location
EAF	Expeditionary Aerospace Force
ENMCS	Engine not mission capable for supply
FMC	Fully mission capable
FOL	Forward operating location
FSL	Forward support location
GUI	Graphical User Interface
JEIM	Jet Engine Intermediate Maintenance [shop]
LANTIRN	Low-Altitude Navigation and Targeting Infrared for Night
MAJCOM	Major command

MDS	Mission design series
MEFPAK	Manning and Equipment Force Packaging
MTW	Major Theater War
NCO	Noncommissioned officer
NMC	Not mission capable
PAA	Primary authorized aircraft
RCM	Reliability-centered maintenance
REMIS	Reliability and Maintainability Information System
RR	Removal rate
SER	Scheduled engine removal
ST	Short tons
TCTO	Time change technical order
UER	Unscheduled engine removal
USAFE	U.S. Air Forces Europe
UTC	Unit Type Code
UTE	Utilization
WRE	War Reserve Engine

Chapter One

ALTERNATIVE JET ENGINE REPAIR STRUCTURES AND THE EXPEDITIONARY AEROSPACE FORCE

THE EXPEDITIONARY AEROSPACE FORCE AND THE CHALLENGE FOR COMBAT SUPPORT

Since the end of the Cold War, the United States has found itself in a new security environment: Instead of facing a known enemy in a limited number of locations (Europe and Korea), the U.S. military has been required to perform numerous overseas deployments—many on short notice—in support of crises ranging from humanitarian relief to Operation Desert Storm. This pattern of fast-breaking, varied regional crises appears to be the model for the foreseeable future. The U.S. Air Force has been and will continue to be heavily involved in all of these operations.

This new environment has placed a substantial burden on the Air Force in personnel as well as equipment. When operations have required land-based airpower, for example, the only option has been to deploy both personnel and equipment to remote locations and keep them there. This combination of frequent and lengthy deployments has created professional and personal turbulence for Air Force personnel and has been linked by some to recent decreases in both retention and readiness.[1]

---

[1]See Richter (1998). However, there is evidence that some deployments help retention (Hosek and Totten, 1998).

1

To ease this burden, the Air Force has begun to reorganize itself into an Expeditionary Aerospace Force (EAF). The main thrust of this reorganization is to replace the forward presence of airpower with a force that can deploy quickly from the continental United States (CONUS) in response to a crisis; commence operations immediately on arrival; and sustain those operations as needed. To implement this vision, the Air Force will divide its forces into roughly ten Aerospace Expeditionary Forces (AEFs), each with a mix of fighters, bombers, and tankers, and will assign two of these forces to be "on call" for crises for 90-day periods, leaving 12 months between on-call periods for each AEF. The on-call AEFs will provide the forces and personnel to staff current rotations such as Operations Northern and Southern Watch[2] on the same 90-day cycle. These arrangements should greatly reduce current turbulence.[3]

The EAF Challenge to the Agile Combat Support/Mobility System

Early discussions of EAF implementation issues largely assumed that after some reengineering, support resources such as munitions, maintenance, and fuels would also be dispatched from CONUS to the deploying unit. Airpower would then be "light, lethal, and self-contained"—or at least self-sustaining for the first few days. However, the current combat support system, like the current combat force, was designed for the scenarios of the old security structure: the two-major-theater-war (MTW) scenarios of the Cold War, in which air units deployed into bases with full support infrastructure in place.[4] In these scenarios, support processes (especially related equipment) were assumed and designed to be in place, and deployability was secondary to capability and cost. These

----

[2]These operations involve the enforcement of no-fly zones over Iraq.

[3]Press conference, August 4, 1998, at the Pentagon, held by then–Acting Secretary of the Air Force F. Whitten Peters and Air Force Chief of Staff General Michael Ryan. See also, Ryan (1998). This is the most comprehensive of several talks on the subject by General Ryan. For a more complete description of the EAF concept and its history, see Davis (1998).

[4]Although not fought in a planned theater, Operation Desert Storm largely fit into this model because of the sophisticated Saudi infrastructure used by the U.S.-led coalition and because of the extensive time the coalition was allowed to enhance and extend that infrastructure.

processes depend on significant numbers of vehicles, material-handling equipment such as forklifts and trailers, bomb-building equipment, fuel-pumping systems, and the like.

In the current environment, the combat support system[5] must support an expanded range of operations from MTWs to small-scale contingencies. Further, these operations may take place in any of a variety of locations, vastly increasing the uncertainty of planning. This significant change in environment means that the system will need to be substantially modified to cover the new range of operations.

Forward Support Locations and Prepositioning

In 1997, RAND began a series of studies at the request of AF/XO and the Air Force Deputy Chief for Installations and Logistics (AF/IL) on the issues raised by the need to adapt combat support to the expeditionary concept. In these studies, RAND researchers developed a framework for the analysis of combat support requirements as a function of operational needs;[6] laid out a vision for strategic combat support planning;[7] and applied the framework and vision to a number of key combat support commodities, such as fuel, munitions, Low-Altitude Navigation and Targeting Infrared for Night (LANTIRN) pod maintenance,[8] and F-15 avionics maintenance.[9]

These studies have made clear the broad characteristics of the combat support system that is required to support expeditionary operations for the current force. The most important finding is that the Air Force's original goal of deploying a complete package of combat

---

[5]Before the development of the EAF concept, the Air Force referred to its combat support system as Agile Combat Support (ACS). With the EAF, the prevailing usage is Expeditionary Combat Support. In this document, we will simply use the term *combat support* to include maintenance, supply, and all other activities that were subsumed under ACS. Since transportation and mobility will be key components of combat support in the expeditionary environment, we will include these functions in the term as well.

[6]See Galway et al. (2000).

[7]See Tripp et al. (1999) and Tripp et al. (2000).

[8]See Feinberg et al. (2001).

[9]See Peltz et al. (2000).

aircraft and support within 48 hours to an unprepared ("bare-base") forward operating location (FOL) cannot be met with today's support processes. That time line can be met only with judicious prepositioning of materiel at FOLs and with the establishment of forward support locations (FSLs) for the storage and maintenance of selected commodities. Support can be provided for heavy combat units from CONUS only if one accepts a time line on the order of a week or more.

This does not mean that expeditionary operations are infeasible. Rather, it means that setting up a strategic support infrastructure to perform such operations involves a series of complicated trade-offs. For example, expensive 48-hour bases with substantial stocks of prepositioned equipment should be reserved for areas of the world that are critical to U.S. interests and are under serious threat, such as Southwest Asia or Korea. In other areas, a 144-hour response may be adequate, and such a deployment could be supported almost completely from suitably chosen and stocked FSLs. In still other areas, such as Central America, the major type of operation contemplated may be humanitarian relief, which could be done with a 48-hour time line to a bare base because much of the heavy equipment required for combat operations would be unnecessary.[10]

Although the research cited above has shown for several different support processes that FSLs and judicious prepositioning are key strategies in implementing the EAF, the analysis needs to be extended to other critical support processes to determine where they should be located. One of those processes is the Jet Engine Intermediate Maintenance shop (JEIM), which provides combat units with extensive repair of jet engines.

THE JET ENGINE INTERMEDIATE MAINTENANCE SHOP

As with much other Air Force maintenance, jet engines—especially those that power fighter aircraft—are repaired largely with a three-level maintenance concept.

---

[10]For a more detailed discussion of the analysis, see Galway et al. (2000). For a more complete picture of the system, see Tripp et al. (2000) and Killingsworth et al. (2000).

1. Flight-line maintenance consists primarily of inspections, diag-
   nostics, and some quick repairs that do not involve engine teardown.

2. Intermediate maintenance at the JEIM involves disassembly of the
   engine; substantial repairs to parts such as the fan, low-
   pressure turbine, and afterburner (in engines so equipped); and
   test-cell runs.

3. Depot maintenance involves complete teardown and refurbish-
   ment of any repairable part in the engine. The engine's usage parameters (flight time, cycles, etc.) can be effectively set to zero by
   such a rebuilding of the engine.

The JEIM at a wing is staffed by the propulsion flight (usually a part
of the Component Repair Squadron). This organization is usually
quite large (100 to 150 people for a fighter wing) and occupies a substantial industrial space equipped with five or more work bays of
1500 square feet each, with an overhead crane, supply storage, back
shops for specialized repair activities, and an off-site test cell in a
"hush house" where a fully assembled engine can be run at full
power for testing purposes.

Traditionally, the JEIM has been located at the operating base with
the aircraft, under the overall command of the operational commander.[11] As with most other maintenance functions, this arrangement
has stemmed from the notion that the operational commander
should have control over all of the resources needed to generate required sorties and that the unit should be relatively self-sufficient in
combat and combat-support capability for a period of weeks. This
concept has been reinforced by prior planning for major wars in
Europe and Korea, where a unit would be moved into theater to
existing bases for immediate action but could expect little resupply
during the first few weeks of combat owing to the need to move in
other forces.

In practice, this collocation of the JEIM with the fighter squadrons
has resulted in a slight blurring of the functions of the two lower

---

[11]For highly reliable engines—especially those in transport aircraft—that spend large
amounts of time away from their home bases, the JEIM has sometimes been located in
a centralized regional facility.

repair levels. The JEIM serves as a source of expertise to back up the flight line, which may have less experienced crews supervised by a few senior engine personnel. It can also provide quick-response repair or cannibalization as necessary for key parts needed by the flight line. In planning for wing deployment, the JEIM is therefore planned to move along with the rest of the wing support—although not with the combat aircraft themselves, who will use spares to replace engines until the JEIM arrives and is up and running. This concept is ingrained in the authorized staffing levels, in the division of experience and responsibility, and even in the information systems supporting engine repair in that only personnel from the wing JEIM are authorized to sign off on JEIM-level repair work for the wing's engines.

As with maintenance concepts for other commodities, this concept is called into question by expeditionary operations. Given the need to move quickly and keep deployment transportation requirements down, the deployment of the JEIM (along with other maintenance back-shop operations) is being reevaluated.

CENTRALIZATION

Even before the reexamination of engine maintenance was made necessary by expeditionary operations, the centralization of JEIM operations had been an ongoing issue for the Air Force engine community. Among the factors favoring centralization was the increased technical complexity of engines—especially high-performance fighter engines, whose exotic materials and extremely demanding tolerances tended to restrict the repairs that could be done in the field even by skilled JEIM technicians. The large investment in facilities, potential gains from economies of scale in the operation of a large repair facility, and changes in experience levels as the armed forces have contracted have all been used as further arguments for centralizing engine repair.[12]

---

[12]RAND researchers, among others, contributed a number of studies in the late 1970s and 1980s on the centralization of different support processes. See, for example, Berman et al. (1975) or Carrillo and Pyles (1982).

Operating in the opposite direction was the fact that, unlike commodities such as avionics components, engines are heavy and bulky, requiring special packing to ship. Current fighter engines, for example, are 16 feet long, weigh 3000 to 4000 pounds, and require several hours of preparation for shipping (the engine fluids need to be drained and purged, etc.). Fighter engines have also been subject to numerous time change technical orders (TCTOs), some of which require immediate, labor-intensive attention to fix safety-of-flight problems. It has been argued that removing, packing, and shipping a large number of engines to a central facility is infeasible in such cases. Of course, the issue of control over maintenance assets has also been a central concern, although it often remains unvoiced in discussions of centralization.

The result of this controversy has been a history of partial centralization of JEIM operations in certain regions and for certain engine types, alternating with a restoration of JEIM facilities to operating units. In the early 1990s, for example, the JEIM maintenance of the Pratt & Whitney F100-220 was moved from the units to a special JEIM facility at San Antonio Air Logistics Center under the control of the Air Force Logistics Command (later the Air Force Materiel Command). This move was strongly opposed by the units, and the experiment was terminated within two years, with JEIMs being restored to the units. Despite this experience, the decline of some fleets and continuing problems in retaining experienced manpower led to other centralizations, most notably the following:

- Centralization of General Electric F110 engines at Misawa, Japan, in support of aircraft at Kunsan, Osan, and Misawa Air Bases.

- Centralization of the B-1B engine JEIM at Dyess and McConnell Air Force Bases.

- Retention of the TF-34 JEIM at Shaw Air Force Base supporting A-10 operations at Pope Air Force Base and Spangdahlem Air Force Base in Germany, even after all A-10s were removed from Shaw.

- Various Air National Guard (ANG) centralized facilities to more efficiently use limited manpower.

In the late 1990s, studies by RAND and others indicated,[13] as noted above, the difficulties involved in attempting to move a complete air base, including the JEIM, to a bare base within a very short time frame. As with other commodities with complex maintenance requirements, this generated renewed interest in centralization.

Finally, in Operation Noble Anvil (the U.S. Air Force part of the operations in Kosovo), logistics planners and operators established centralized engine repair facilities at established European bases to support forces flowing into bases that either were newly established (in southern Italy) or had limited facilities (e.g., Aviano). The JEIM at Lakenheath Air Force Base supported F-15Es from Lakenheath as they were deployed several times to other bases during the operation, and Spangdahlem restarted TF-34 repair to support active and Guard A-10s operating in Italy while preparing to increase F100 repair if needed.

The recent partial centralizations, the experience in Kosovo, and the continuing pressure from expeditionary operation concepts to reduce deployed footprint prompted the Engine Executive Advisory Group, at its spring 1999 meeting, to call for a fresh examination of engine repair centralization. Air Combat Command (ACC/XR)[14] took the lead in answering this question and in May 1999 asked RAND to extend the analyses done on expeditionary support for other commodities to the JEIM function.

SCOPE OF THE PROJECT AND REPORT

Combining the concerns of the Engine Executive Advisory Group and the overall thrusts of RAND EAF support research, this project sought to evaluate several different alternatives for accomplishing JEIM support. The two fundamental alternatives for JEIM support are to provide intermediate repair at the supported base (decentralized) or to provide such support at a centralized, off-base facility. In peacetime,

---

[13]See Galway et al. (2000) and Tripp et al. (2000). Various analyses by AF/ILXX and their contractors were also influential and were briefed to AF/IL.

[14]At this time, ACC/XR had responsibility for weapon system support, with other logistics responsibilities being exercised by ACC/LG. In 2000, the ACC staff was reorganized so that all logistics functions were returned to LG, in line with staff organization at other major commands.

the centralized facility would be in CONUS, but during a conflict the facility could also be located in the theater of operations. In Chapter Three, we will enumerate and define these alternatives more precisely in terms of the specific assumptions made for each.

In evaluating the performance of the alternatives, three broad sets of metrics must be considered. The first is performance: Does the alternative provide the required support for operational flying? In peacetime, this means being able to maintain the requisite flying for pilot training; in wartime it means being able to meet the required sorties day by day as the conflict proceeds.

The second set of metrics speaks to resources: What does the alternative require to provide adequate performance? For jet engines, one of the key resources is spare engines. Over some time period these are fungible with repair, but they also provide a hedge against uncertainty, and we shall require the alternatives to be able to maintain a spares level to meet surprises. The other resources are personnel and transportation. Centralized facilities can often reduce personnel requirements, but at the expense of requiring transportation. This evaluation should give us an indication of the trade-off between these two elements. All of these resources will contribute to peacetime costs, which must be considered at least implicitly in any evaluation of support alternatives.

The final set of metrics for each support alternative addresses how well that alternative responds to unforeseen events, such as changes in scenario, transportation times, or removal rates.

In Chapter Two, we describe the analytic methodology of our study. To evaluate the alternatives, we decided on a simulation modeling approach; accordingly, we discuss the rationale for that decision, the structure of the model, and the ancillary data analyses that we used to determine model parameters and to assess current repair alternatives. Chapter Three gives the results of our analysis. After describing the alternatives in detail, we evaluate the performance of each alternative in an MTW scenario for each of the three engines we studied: The Pratt & Whitney F100-220 and F100-229 (F-15s and F-16s) and the General Electric TF-34 (A-10). Chapter Four expands the analysis in some selected directions to assess the robustness of our results.   Finally, Chapter Five gives our conclusions and recommendations.

Chapter Two

ANALYSIS METHODOLOGY

Our method of studying alternative concepts of JEIM support rested on the development and use of a simulation model of the JEIM and related systems, supported by ancillary data analyses both to provide input to the model and to answer other questions not directly addressed by that model. In this chapter, we give our rationale for developing a fairly complex simulation model, describe the essentials of the model itself, and discuss our data sources for the other analyses.

RATIONALE FOR DEVELOPING A SIMULATION MODEL

As we stated in the first section, our aim in this project was to compare several different alternatives to the current support concept of decentralized JEIM. These alternatives included full centralization in peace and war and several hybrid systems (e.g., decentralized in peace but centralized in the theater of operations). We must compare these alternatives using several different performance metrics and potentially over different scenarios as well. There are two contrasting approaches toward doing this analysis:

- Use data from the previous history of centralization attempts to determine whether or not centralization will work.

- Develop a model of the JEIM and supporting systems such as transportation, and evaluate the alternatives with the model.

Our judgment was that the history of centralization was of limited utility in the basic assessment of JEIM alternatives, although it did provide much material to help shape our choice of alternatives and information on key factors that caused problems in centralized repair. For example, we wanted to be able to examine the effects of centralizing intermediate repair for engines that had never had centralized repair (e.g., the F100-229); to look at full centralization of engines that had had only partially centralized repair (e.g., the TF-34); and to look at engines for which centralization had failed (e.g., the F100-220). In addition, many of the centralization efforts, both successful and unsuccessful, had often had specific features, such as location, driven by external constraints that may not apply in general situations.

Finally, the data available on pre- and postcentralization performance were limited. In particular, there was no information on any of the major centralized facilities during a conflict, since few such facilities have supported a conflict as the major source of repair.[1]

We therefore turned to modeling as our primary tool for this study. Most previous RAND work on supporting the EAF[2] had used deterministic models, which use the means of stochastic quantities (such as transportation times) in analytic formulas. The goals of our study, however, led us instead to build a discrete event simulation model in which individual engines were represented by identifiable entities and stochastic elements are better represented.

The first and most important reason for simulation was that the metrics in which we are interested, such as sorties missed, current spares levels, and queue sizes at key shop points, are inherently dynamic—that is to say, we are often interested in their pattern over the course of time rather than simply an overall average. In particular, in conflict situations where sortie requirements may change over time (see the discussion in Chapter Three about the sorties required in our notional MTW scenario), we want to see the

---

[1]The TF-34 centralized JEIM did do repair work for deployed forces during Kosovo, but some capability was established at Spangdahlem as well. The Lakenheath JEIM supported its own fighters in deployed locations but did not repair engines from other units.

[2]See Feinberg et al. (2001) and Peltz et al. (2000).

value of key metrics day by day. A force may miss only 5 percent of the sorties required of them, but there is a big difference in performance if those 5 percent constitute 20 percent of the sorties required during the first few days of a war rather than 5 percent at the end of a conflict. Simulation allows us to look at dynamic metrics at the level of resolution of the model.

Second, by modeling the engine repair system directly, we are able to impose and observe the effects of capacity constraints directly in much the same way as if we were observing the real system (subject, of course, to the simplifications built into the model).

Third, by using a discrete-event, closed-loop simulation with a fixed set of engines that do not enter or leave the model, we ensure that there is a realistic relationship between repair capability and engine usage: If an engine shop is able to generate (fix) more engines, then a fewer number of aircraft will have holes. This will in turn mean that fewer sorties will be missed owing to engines and therefore that a greater number of sorties will be flown. The increase in utilization may increase the number of engines that fail, which will in turn put pressure on the engine shop and reduce its production rate. The opposite is also true: Reduced repair ability reduces the number of engine failures.

Finally, key management decisions in engine repair are based on the characteristics of the engine in work. For example, engines for engaged forces are given priority during a conflict.

In addition to these considerations, the development of the model allowed us to lay the foundation for future expanded studies of engine repair by incorporating other important characteristics, such as the management of engine deployment and repair based on the time characteristics of individual engines; the effects of engine demographics[3] and different management decisions on JEIM and depot workload; a more detailed representation of repair modes based on whether an engine removal is scheduled (for an inspection or to

---

[3]Demographics refers to the age distribution in terms of parameters such as cumulative flying hours; these drive the inspection and removal of many critical components of the engine. Depot repair, as noted above, usually "zero-times" the engine. The distribution of ages at a particular point in time is an important determinant of JEIM (and depot) workload; conversely, modifying workload can manipulate age distribution.

change a part that has reached a specific age) or unscheduled (owing to a malfunction of some type); and transportation policies. With our simulation model, these extensions can be easily and naturally added in the future.

A number of simulation models that treat repair systems have been used for analyses similar to ours, the most notable of which is Dyna-METRIC, a model that was developed by RAND in the 1980s and was used extensively in various versions for a number of studies during the 1980s and early 1990s. Even the most recent versions of Dyna-METRIC, however, do not satisfy the requirements outlined above: It does not track individual units with specific properties, and it does not have much detail in its representation of transportation. Further, it is written in FORTRAN and is difficult to modify internally to handle some of these potential extensions.

Many of these drawbacks are not present in current graphical user interface (GUI) simulation packages. These packages, which draw on progress made in programming languages, user interfaces, and hardware capabilities, make it possible to quickly design, write, and use simulations whose complexity and detail would have been impossible with computing resources available only a decade ago. Some initial experimentation indicated that these packages could indeed provide us with a simulation that ran in reasonable times when simulating fleets of the sizes encountered in fighter engines. This led us to build the model described below, even though it required a substantial investment in initial effort, as a flexible tool that could be used for this study and for future investigations.

MODEL DESCRIPTION

In performing the analysis described in the chapters that follow, we developed a suite of simulation models with Extend, a GUI-based system and process modeling software package.[4] In this chapter, we give an overview of the basic core structure of the models and the ways in which that core structure was extended and modified to model different engine repair alternatives and scenarios.[5]

---

[4]Extend is created by Imagine That, Inc.

[5]For a complete description of the models, see Amouzegar and Galway (forthcoming).

Overall Structure of Jet Engine Repair

The usage and maintenance of jet engines comprises the sequence of events illustrated in Figure 2.1: Planes fly sorties from bases in peacetime to meet training requirements and from FOLs in wartime to accomplish combat missions. After each mission, the planes' engines are inspected on the flight line and, depending on the accumulated flying hours,[6] are given minor maintenance. Engines may also be removed from aircraft and sent to a JEIM facility for scheduled major maintenance—e.g., an inspection or scheduled

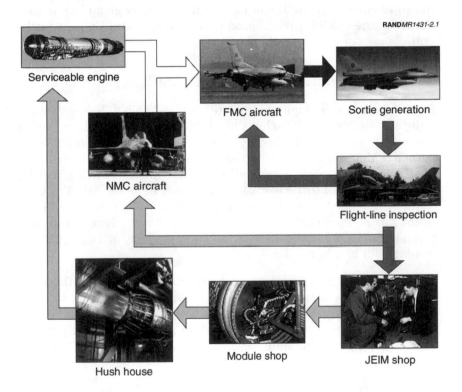

RAND*MR1431-2.1*

Serviceable engine

FMC aircraft

Sortie generation

NMC aircraft

Flight-line inspection

Hush house

Module shop

JEIM shop

Figure 2.1—Operation and Maintenance Sequence

---

[6]The model also keeps track of engine serial numbers and aircraft tail numbers throughout the simulation. Each engine is linked to a particular tail number until it is removed for maintenance.

replacement of parts that are time-limited—or for unscheduled maintenance, such as that made necessary by foreign object damage. At the JEIM, the engines are inspected, repaired, tested by being run in a test cell and are then returned to the flight line as serviceable spares. Bases and FOLs use spare engines to replace engines sent to the JEIM, but they may be forced to miss sorties when the spares are exhausted if the aircraft with serviceable engines cannot be used for multiple sorties.

Our models represented each of these processes as a series of queues and delays based on data collected from units and Air Force information systems. More detail on our modeling is given below (full details of the model will be found in Amouzegar and Galway, forthcoming).

Sortie Generation and Flight-Line Operations

At each base, sortie requirements are imposed daily.[7] For each set of daily sorties, the model selects available aircraft with engines that have the fewest accumulated flying hours. Aircraft availability is assumed to depend solely on engine availability. If too few aircraft are available owing to lack of engines, then missed sorties will be recorded. Once a sortie is missed, it cannot be made up. For each sortie, the engine's usage is incremented to reflect the number of hours flown. If an engine has accumulated enough flying hours, it is given minor maintenance, which represents a delay of a day until it is available for another sortie.

An engine may, however, have to be removed and sent to the JEIM either because it is due for scheduled maintenance or because it has suffered some type of failure, such as foreign object damage. In our model, the probability of removal is determined by multiplying the removal rate per flying hour for a specific engine type by the length

---

[7]In peacetime, the daily sortie requirement is computed by multiplying the number of aircraft at the base (the primary authorized aircraft, or PAA) by the utilization (UTE) rate (the number of times each aircraft is to be flown each month divided by 30). In wartime, the number of sorties per day is specified directly.

of the completed sortie.[8] The removal rates we used were taken from reports produced by the Air Logistics Center managing the engine.[9]

If a spare engine is available at the base, the spare is installed in the aircraft and it is returned to the pool of available aircraft. If no spare is available, the aircraft cannot be used for sorties. If the aircraft had a second engine that was not sent to the JEIM (as is the case with F-15s and A-10s), that engine is counted as a serviceable spare (i.e., the base practices engine cannibalization to the fullest extent).[10]

Transportation

When the JEIM facility is located on the base from which the aircraft are flying, engines enter it immediately on removal, but when engines are sent to a remote centralized facility, they must be transported via air or ground. In our analysis, we assume that engines are shipped as soon as they are designated for the JEIM and that the only difference between transportation within CONUS, within a theater, or between CONUS and theater is the time required to do the movement (engine preparation time is part of this time). We have thus assumed the existence of a responsive transportation system when repair is centralized.[11]

Jet Engine Intermediate Maintenance

When an engine reaches the JEIM facility, it enters a queue for inspection, which takes one to two days. After inspection, it may be

---

[8]Technically, a Bernoulli random variable is generated with the specified probability of one (removal). If it comes up one, the engine is removed and sent to the JEIM; if it comes up zero, there is no failure and the engine stays on the aircraft.

[9]We had initially planned to treat scheduled and unscheduled removals separately, with scheduled removals being driven directly by engine usage. However, we discovered that the available data on the repair process were not sufficient for us to determine a difference in repair times. In many cases the reported data for scheduled and unscheduled maintenance tasks were combined.

[10]Engine cannibalization is a time-consuming process that the flight line would try to avoid in peacetime. In a conflict, the overriding goal is to keep aircraft fully mission capable (FMC), so cannibalization would be more aggressive.

[11]See Peltz et al. (2000), Feinberg et al. (2001), and Tripp et al. (2000) for a discussion of the need for reliable transportation in centralized repair.

found that the required parts are not available, and the engine will enter the status of ENMCS (engine not mission capable for supply). In our model, the probability of going ENMCS and the time engines spend in that state are based on empirical data from the Comprehensive Engine Management System (CEMS).[12]

Once all of the needed parts are on hand,[13] the JEIM personnel repair it provided that capacity is available. Capacity consists of labor together with physical equipment (rails and tools) and is denoted in our analysis by the term *rail team*; the capacity for a JEIM is one of its key characteristics.[14] If no rail team is available, the engine waits in a queue until a rail team is free. For modular engines such as the F100-229 and the F100-220, some modules are repaired in the module shop, a separate area of the JEIM where problems with individual modules are worked in parallel.[15] Each of these repairs is represented by a delay and, like the repairs in the main JEIM, is subject to the availability of labor and capacity. For both main shop repair and module repair, delays are randomly drawn from an empirical distribution taken from CEMS data. At the end of repair, the engine is reassembled. Like the other repair processes, the assembly process is modeled as a delay. When assembly is completed, each engine flows to a queue for the test cells.

Test cells are limited resources, with most bases having only a single cell that can test only one engine at a time. The test takes one day, except for a small proportion of engines that fail initially and require an extra day of tests (this fraction is based on our unit interviews).

---

[12]Our analysis of the data indicates that only a fraction of engines go ENMCS, and of those that do, the duration of ENMCS can last from a few days to months. In the model, we also select a fraction of engines to go ENMCS; the duration of the ENMCS is randomly drawn from the empirical distribution of ENMCS in the CEMS data.

[13]This is not strictly realistic in that an engine may have several episodes of ENMCS status during a single repair as the extent of repair is determined. However, the overall repair time is still faithfully represented by separating the total wait for parts and the total working time into two sequential delays.

[14]The engines are mounted on structures called rails for repair. A rail team is defined as a minimum number of personnel needed to work on an engine in a two-shift day. For example, for 229 engines, a rail team consists of five people per shift, or a total of ten people.

[15]While it is possible for a module to be awaiting parts, we do not have detailed information on the modules. These delays are subsumed in the ENMCS distribution for the entire engine.

When the engines complete their testing, they move to final inspection and are then returned as FMC engines to their respective bases and FOLs, where they serve as serviceable spares and are installed in aircraft as needed.

DATA SOURCES

Our modeling was supplemented by analysis of data drawn from the CEMS; from the Reliability and Maintainability Information System (REMIS), which rolls up data from the base-level Core Automated Maintenance System (CAMS); and from data provided in both electronic and paper form by the units we visited.

The CEMS data provided information on total repair time for individual engines, ENMCS times, and transportation times for engines such as the TF-34, for which Shaw Air Force Base provides JEIM repairs for some operational bases. REMIS provided a check on the CEMS data for overall engine repair and provided repair data for module work. However, neither could easily give us information linking module work to specific engines. REMIS has space for the engine serial number in the module repair records, but the field is seldom used. CEMS has recently started tracking module repair, and data series sufficient for analysis will be available in a matter of years. We attempted to get overall counts from REMIS of module repairs per engine inducted into the JEIM, but these were eventually deemed unreliable because it was difficult to distinguish between scheduled and unscheduled work (many jobs are a mix, and the job is often coded as unscheduled work when it is started). For these reasons, module repair is an area of our modeling that requires more work.

Finally, we did use interviews at the units, initiated with our REMIS and CEMS analyses, to better understand the details of repair.

Chapter Three

ASSESSMENTS OF ALTERNATIVE STRUCTURES
FOR JEIM SUPPORT

OVERVIEW

In this chapter, we first present a detailed description of the alternatives we considered and our evaluation design. For each engine, we then evaluate the performance of each alternative during an MTW scenario. A separate analysis of the performance of relevant centralized and decentralized alternatives in peacetime is provided in the appendix.

REPAIR ALTERNATIVES

In Chapter Two, we discussed the general concepts of decentralized and centralized repair. Here we precisely define and describe in detail the specific JEIM alternatives that we evaluate in this analysis.[1]

Decentralized-deployed (DecDep). In this alternative, which is the current plan for deployed engine support, peacetime maintenance is provided by JEIMs located at each base. When part of a unit is deployed, part of that unit's JEIM deploys to the appropriate FOL to

---

[1]We label each alternative in terms of "peacetime repair–wartime repair." For example, the decentralized-deployed case implies a decentralized mode of repair during peacetime (and for nonengaged forces) at home units and deployed JEIM shops at the FOLs. Following the name of each alternative, we give a short abbreviation in parentheses that will be used to label the appropriate graphical results later in the chapter.

form a deployed JEIM. As per current plans, the JEIM deploys by day 30 of the war and begins working immediately, but the test cell is not ready to test repaired engines until day 60.[2] The transportation requirement for this alternative is that needed to deploy the JEIM itself.

Decentralized–no deployment (Home). As with the previous alternative, each of the peacetime bases has its own JEIM, but the home JEIM supports any deployed forces from its unit as well.[3] The home JEIM is sized so that it has the resources to support both peacetime and wartime flying. This alternative requires intertheater transportation to move engines between the FOL and the home base.

Decentralized–forward support location (FSL). As with the previous two alternatives, each peacetime base has its own JEIM. When the units deploy, however, some JEIM personnel (but not their equipment) deploy to a single FSL in theater from which all deployed units are supported. We assume that the FSL is "lukewarm"—i.e., that it is ready to begin operations as soon as the JEIM personnel arrive. In this case, there is no additional delay for the test-cell setup, but there may be some delay for the arrival of personnel. During the conflict, intratheater transportation is required to move engines between the FOL and the FSL.

CONUS support location–forward support location (CSL-FSL). In this alternative, all units are supported in peacetime by a CSL that deploys personnel to an FSL in theater when conflict occurs. We must staff the CSL in peacetime with the sum of the rail teams needed for deployment and those required to keep the nonengaged forces flying. (Note that for deployed forces, this alternative is indistinguishable from the previous one, since the repair structure in theater is identical.) For this alternative, the conflict situation requires intratheater transportation identical to that of the previous case.

---

[2]This limitation is due to the requirement that the test-cell foundation be strong enough to resist the thrust of modern fighter engines at full military power (afterburner—about 29,000 pounds of thrust for the F100-229). The foundation is a concrete slab that must set for 30 days after pouring.

[3]Note that some units use this method today to support their deployments to Operations Northern Watch and Southern Watch (the enforcement of Iraqi no-fly zones).

CONUS support location (CSL). In this last alternative, all units everywhere are supported by a single CSL both during peacetime and in deployment. During the conflict, engines must be moved by intertheater lift from the FOLs to the CSL.

EVALUATION DESIGN

Maintenance alternatives are evaluated through a comparison of their performance (how well they support the forces) and according to the resources they require. In this study, we use the following two primary performance metrics:

- **Percentage of missed sorties.**[4] This is the fundamental metric for any logistics process: Can it provide enough operational aircraft to accomplish the required sorties? In peacetime, these are training sorties that are conducted to allow pilots to improve and maintain proficiency; in war, they are the combat sorties (which may vary daily with the phase of the conflict) that are required to execute the operations planned by the commander.

- **Spares levels.** The use of spares levels as a metric may be somewhat surprising, since spares are a resource like repair capacity (i.e., we can dispense with repair if there are sufficient spares and vice versa). However, units rate their engine repair performance by the number of serviceable spares they have on hand; although units can avoid missed sorties even with holes in aircraft due to engines in repair, a positive spares level (i.e., no aircraft missing an engine) provides a hedge against uncertainty. If the unit has a positive spares level, all aircraft can be flown to safety if needed, and there is a buffer available for a sudden surge in demand or other unanticipated events.

The primary resource measure in this analysis is the number of rail teams in the JEIM facilities. All other things being equal, we prefer to get the same or superior performance from smaller numbers of rail teams, particularly in conflicts where we want to hold down the

---

[4]Daily sortie demand (peacetime UTE rate or wartime schedule) is translated into a daily aircraft requirement (each aircraft flying no more than two sorties per day), and the percentage of missed sorties is computed from the ratio of required and available aircraft.

number of troops in harm's way. However, transportation requirements and the risk they entail must also be taken into consideration.

In our evaluation, we give each alternative the number of rail teams required to achieve identical performance on the first metric, missed sorties: We require that on average, no wartime sorties be missed over the entire course of the conflict (within simulation error). (We will always add enough repair resources in CONUS to ensure that aircraft left behind have enough engine support to allow regular peacetime training missions to be flown.) We then rank the alternatives by their performance on average daily spares levels during the conflict and by the rail teams required.

This process leaves several resources and key parameters free, including the following:

- The engine removal rate.

- The repair time (time actually in work and time spent ENMCS).

- The total number of spares and the number of these that are war reserve engines (WRE).[5]

- For the centralized alternatives, the transportation time to JEIM facilities.

Because all but the last parameter are currently "fixed" by policy, current technology, and current practice, in our analyses they will be set to current values unless otherwise noted. The engine removal rate was obtained from reports on engine removals from San Antonio Air Logistics Center. The repair time and ENMCS time are empirical statistical distributions constructed from unit interviews and CEMS data. The number of spares and the subset that are WRE were extracted from briefings prepared by ACC/LGP on engine and fleet readiness, and in all cases spares were located at the bases from which the aircraft flew. We used the current engine fleet sizes because there are no plans to buy further spare engines for the engines we analyzed.

---

[5]Only a portion of the spare engines are deployed with the engaged forces (these spares are designated as WRE). Unless otherwise noted, we use the current WRE when we refer to spare engines deployed.

In this chapter, we have used values for transportation times that we consider plausible according to current practice:

- Within CONUS, two to four days.
- Between CONUS and the theater of conflict, 15 days.[6]
- Within the theater of conflict, two days.

In Chapter Four, we will perform sensitivity analyses with selected parameters to determine the robustness of our results if those parameters should vary.

We use the single-MTW scenario in this analysis because this is the most stressing plausible scenario for engine repair. Unless otherwise noted, we run the models for two simulated years. The conflict begins after one year of peace and ends in 100 days, after which all units return home to their bases and resume a peacetime schedule. This allows adequate time for the repair facilities to achieve a steady load before the conflict begins.

In all cases, all JEIMs go to wartime work rules (operation in two 12-hour shifts, seven days a week) when the conflict starts, resulting in a proportionate reduction in the time engines spent in work in the JEIM shop (due to the increased work time of the longer shifts). The ENMCS time is also reduced by 40 percent to represent increased levels of parts support by engaged forces during a conflict.[7] For those alternatives in which a JEIM is supporting both engaged and nonengaged units, the engaged units get first priority on repairs.

---

[6]This figure has elicited substantial discussion when briefed to various audiences. Some argue that current transportation times to locations outside CONUS can be substantially longer than 15 days, particularly for large items such as engines. Others argue that planned changes to DoD transportation policies will result in significantly shorter times. Given current constraints on military airlift and given our assumption that this is an MTW (which would likely stress the airlift system), we have elected to retain the 15-day time. Clearly, if this time could be substantially and reliably shortened, locating engine repair in CONUS would become more feasible.

[7]One external reader questioned whether this reduction would be achieved during an MTW.

ANALYSIS OF THE F100-229

For our first analyses, we selected the Pratt & Whitney F100 series because they are modular engines—i.e., each engine is divided into several modules that are designed to be conveniently interchanged in the field. Such modules can be repaired independent of the rest of the engine either at the JEIM or following return to the depot. The use of a JEIM module shop for module repair (unlike the non-modular engines, such as the General Electric F110 series) makes the JEIM model for the F100 the most complex to develop. The JEIM for other engine types can then be derived from the F100 model simply through the elimination of the module shop and through use of parameters appropriate for that engine.

Within the F100 series, we selected the F100-229 because it is the newest version of the F100 operated by the Air Force, having entered service in 1991. As such, it is a fairly small fleet with a total of approximately 260 engines. The correspondingly smaller data set made it easier for us to gain a better understanding of repair dynamics. In addition, the engine currently has no major overhauls due in the near future.

As our removal rate for the F100-229, we use a total removal rate of 5.0 per 1000 engine hours (this combines the scheduled and unscheduled removal rates).[8] On the basis of our data analyses and interviews, we take repair time for the F100-229 (exclusive of ENMCS and delays for labor availability) to be 12 to 21 days in peacetime and 8 to 15 days during war.

During peacetime, the F-15s fly at a utilization (UTE) rate of 18 and the F-16s at a UTE rate of 19.[9] The wartime scenario starts from a peacetime configuration of bases in which most of the F100-229-

---

[8]As mentioned earlier, the model is designed to separately accommodate scheduled and unscheduled repair time. However, not enough reliable data were available to effectively represent the differences. For more information see Amouzegar and Galway (forthcoming).

[9]The UTE rate is the number of times a single aircraft flies in a month; the figure used here is the current ACC target. We note that as of this writing, a number of factors were forcing the actual UTE for ACC units to be less than this. In addition, recent RAND research (Taylor et al., 2000) suggests that even the ACC target may be too low to maintain pilot proficiency and allow newer pilots to acquire needed skills.

equipped aircraft are based in CONUS.[10] At the beginning of the conflict (one year into the model run period), 48 F-15s and 24 F-16s deploy to two single-mission-design-series (MDS) bases in a theater of conflict. The wartime flying schedule has a ten-day surge in which the F-15s fly approximately 1.6 sorties per day and the F-16s about 2.0 sorties per day, followed by a 90-day sustain period in which both the F-15 and F-16 flying schedule is 1.0 sortie per day (see Table 3.1).

In the following analysis of each alternative, we first use the model to determine the number of rail teams (and test cells) that are required to achieve the target of no missed wartime sorties. Using that capacity, we then run the wartime scenario to compute the daily spares profile. Finally, we give the transportation requirements for each alternative.

Table 3.1

Operational Data for Aircraft with F100-229 Engines

| MDS | Aircraft | | Flying Schedule | | | Engine Removal |
	Total	Deployed	UTE	Surge	Sustain	(per 1000 flying hours)
F-15	66	48	18	1.6	1.0	5.0
F-16	36	24	19	2.0	1.0	5.0

Decentralized-Deployed Alternative

The first step is to use missing sorties to size the number of rail teams required to deploy. Figure 3.1 summarizes the model runs[11] by plotting the number of rail teams at each FOL against the percentage of missing sorties for each FOL. From this figure, we can see that 12 rail teams must be deployed—four to the F-16 FOL and eight to the F-15

---

[10]See the appendix for details. We did this because the small size of the force made more realistic basing patterns uninteresting. For engines such as the F100-220 and TF-34, the basing patterns reflect current basing.

[11]Each point on the two curves is the average of several hundred simulation runs, where the final number was selected to make the simulation error of the average negligible.

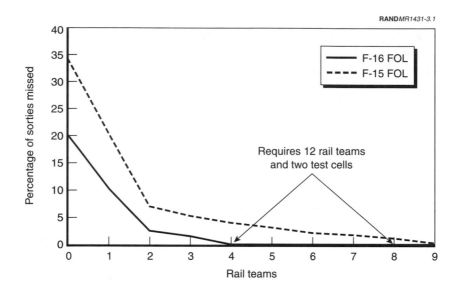

Figure 3.1—F100-229 Decentralized-Deployed Rail Team Requirements

FOL—in order for no wartime sorties to be missed.[12] The model also indicates that each FOL needs one test cell. Together with the four rail teams and four test cells that are retained for the nonengaged forces at the peacetime bases, this alternative requires 16 rail teams and six test cells.

Figure 3.2 shows the daily spares levels achieved by this alternative; the curves are the average of daily spares levels over hundreds of simulation runs. The horizontal axis represents the 100 days of war (recall that the conflict starts on day 365 of the simulation and ends on day 465). The vertical axis represents the serviceable spares, and the dashed lines mark the threshold for the spares levels required to maintain minimum sortie rate (i.e., sorties are lost if the serviceable spares fall below these lines). Negative spares levels indicate aircraft without serviceable spares—e.g., –10 serviceable spares imply that

_____

[12]Although the F-15 line in the figure may justify nine rail teams for the F-15 units, we chose to take the smaller number, since we allow up to 2 percent deviation from any set target due to simulation error.

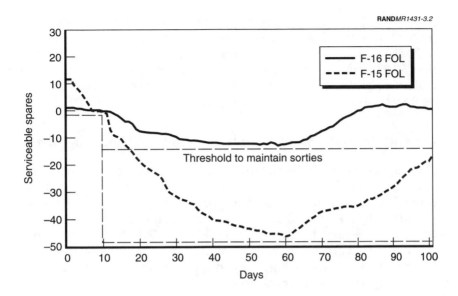

Figure 3.2—F100-229 Decentralized-Deployed Spares Performance

ten F-16s or five F-15s are not mission capable (NMC) owing to engines.

From Figure 3.2, we see that although few sorties are missed in this alternative, the spares performance is very disconcerting. For both bases, the spares level becomes negative by about the end of the surge period and does not begin to recover until the test cell is finally operational at day 60. Unfortunately, deploying more JEIM rail teams cannot solve this problem. In separate runs, the same pattern was observed even when as many as ten times the number of teams were deployed. The problem is that the WRE spares deployed are too few (four WRE for the F-16s and 11 for the F-15s) to cover the initial period when no JEIM is operational. See Chapter Four for a treatment of performance when all spares are deployed as opposed to just the WRE alone.

The transportation requirement for this alternative is that required to move the JEIMs to the FOLs. The latest Air Force Unit Type Code

(UTC)[13] list[14] describes several different F-15 and F-16 independent (capable of operating by themselves) JEIM UTCs that range from 25 to 50 short tons (ST) in weight. Making the conservative assumption that 50 ST must be moved to the F-16 FOL and 100 ST to the F-15 FOL, the airlift requirement is 1.2 C-5 sorties. These sorties need to be flown only once during the first 30 days of the war.[15]

Decentralized–No Deployment Alternative (Home Support)

Once again, we begin by determining the resources required to reduce missing sorties. Figure 3.3 shows the results of this process. It should be noted that since both engaged and nonengaged forces are supported by the home units in this scenario, the results in Figure 3.3 represent the total resources needed in CONUS to maintain both training and wartime missions. Spares performance is shown in Figure 3.4. As before, using only the specified WRE will leave spares levels dangerously low for the deployed forces.

With this alternative, we require intertheater transportation of engines to and from the home JEIMs. For this scenario, an average of 32 engines were returned over the first two weeks (the surge) and ten engines per week during sustainment. This is equivalent to 1.7 C-5s during the first two weeks (each way) and 0.6 C-5 per week for the duration of the war. We have assumed that the engines are moved within a day of the time they are removed or repaired; therefore, these numbers are the gross capacity needed, presumably taken out of an ongoing transportation operation. If the engines are batched to make a bigger load, the performance of this alternative will be worse than that presented here owing to the increased delays.

---

[13]UTCs are sets of equipment and/or personnel required to fulfill different functions in a deployed unit. They are grouped into UTCs as the smallest deployment element; it is characterized by movement requirements in terms of weight, personnel, and the amount of over- and outsized cargo.

[14]Manning and Equipment Force Packaging (MEFPAK) Summary Report, October 27, 1999.

[15]This does not include the resupply transportation (spare parts and modules) required to support the deployed JEIMs.

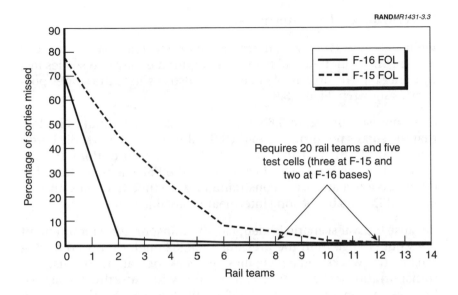

Figure 3.3—F100-229 Home Support Rail Team Requirements

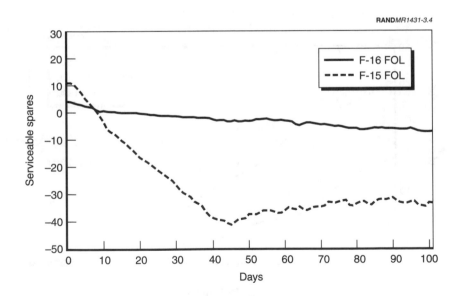

Figure 3.4—F100-229 Home Support Spares Performance

Decentralized-FSL Alternative

Figure 3.5 shows the sizing of the FSL in terms of rail teams deployed to the FSL. Eight total rail teams are required to have no sorties lost by either type of aircraft. As before, we also evaluate the spares posture, as shown in Figure 3.6.

A comparison of Figure 3.6 with Figures 3.2 and 3.4 makes it clear that the spares performance of the FSL alternative is markedly better than that of either of the two fully decentralized options treated above. The key is the speed with which the FSL starts working; the other two alternatives are constrained by the time to deploy and set up the JEIM and by the long intertheater pipeline times.

Because the war scenario is identical, the transportation requirement does not change from the previous case: 32 engines moved during the first two weeks and ten engines per week thereafter. Because this transportation requirement can be met with intratheater assets, however, we express this in C-130 equivalents: 32 for the first two weeks and ten per week thereafter.

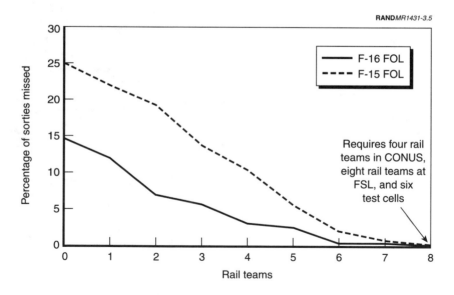

Figure 3.5—F100-229 Decentralized-FSL Rail Team Requirements

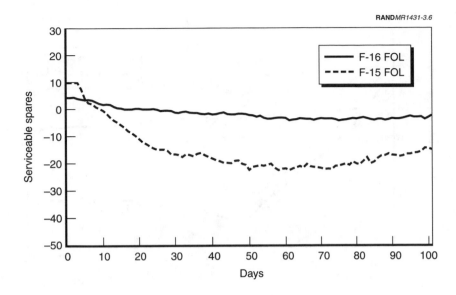

Figure 3.6—F100-229 Decentralized-FSL Spares Performance

CSL-FSL Alternative

As noted above, the requirements for the FSL are the same as those in the previous case. To support the units at home, the CSL requires a residual three rail teams and three test cells (to cover the entire peacetime workload), resulting in 11 rail teams and three test cells. The transportation requirements are identical to those in the previous case.

CSL Alternative

We will omit the sizing analysis and simply report that this alternative requires 12 rail teams and three test cells. As shown in Figure 3.7, however, the spares performance is on a par with the decentralized alternatives.

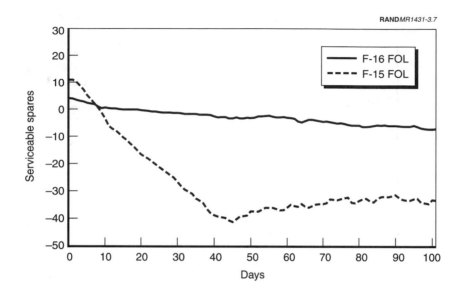

Figure 3.7—F100-229 CSL Spares Performance

Results for the F100-229

The spares results from all four different alternatives are shown in Figures 3.8 and 3.9 (recall that the decentralized-FSL and CSL-FSL alternatives have identical spares performance curves). From the performance results, the FSL looks clearly superior in supporting a fast-breaking conflict, primarily as a result of its fast response. The alternatives of deploying a JEIM to a bare base or supporting engaged forces from CONUS perform less well because of the length of time required before they can start working on broken engines. After the deployed JEIM arrives at an FOL, it can work down the backlog fairly quickly, but the delay until it is ready to produce means that there is a large backlog to work on and a dangerously low level of spares. The CSL and home alternatives, in contrast, are fighting against the long pipeline, which has the same effect as a delay in beginning repair. Increasing spares will help all of the alternatives, but without more spares, only the FSL alternative can use more repair resources.

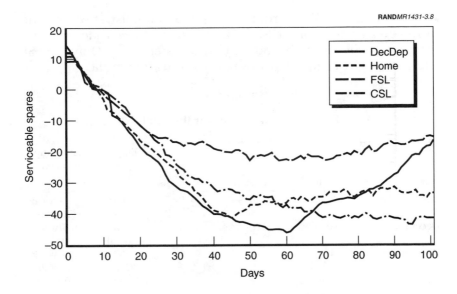

Figure 3.8—F100-229 Spares Performance for All Alternatives for the F-15 FOL

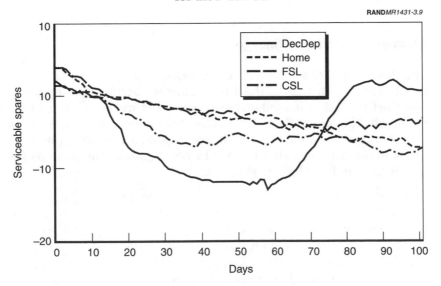

Figure 3.9—F100-229 Spares Performance for All Alternatives for the F-16 FOL

We can summarize the wartime analysis for our illustrative scenario in Table 3.2 below. For the F100-229, the decentralized-deployed alternative is not acceptable owing to the spares levels. Only the FSL alternatives give adequate performance during the conflict.

Table 3.2

Wartime Requirements for the F100-229

Peace	War	Rail Teams	Test Cells	Engine Transportation (aircraft equivalent) 1.2 (C-5)	
Decentralize	Deploy	16	6	Days 1–15	Days 15–100
Decentralize	No deploy	20	5	1.7 (C-5)	0.6/week (C-5)
Decentralize	FSL	12	6	32 (C-130)	10/week (C-130)
CSL	FSL	11	3	32 (C-130)	10/week (C-130)
CSL	FSL	12	3	1.7 (C-5)	0.6/week (C-5)

ANALYSIS OF THE F100-220

The fleet size of the F100-220, the predecessor of the F100-229, is much larger than that of the F100-229, numbering 1200 engines. Like the F100-229, it is used on F-15s and F-16s and is a more mature engine, having entered service in the 1980s. Since the F100-220 is also modular, the model used for the F100-229 can be applied directly to the former with a change in parameter values such as removal rates and ENMCS distribution.

The peacetime total removal rate for the F100-220 currently differs for those installed on F-15s and F-16s. For the F-15, the removal rate is 5.0 per 1000 engine hours (2.5 unscheduled engine removals [UERs] and 2.5 scheduled engine removals [SERs]), whereas for the F-16 it is 7.5 per 1000 engine hours (5.0 UERs and 2.5 SERs). In wartime we use a single removal rate of 5.0 per 1000 hours for both aircraft.[16] We use the same repair time as for the F100-229, consistent with our data analysis and unit interviews. Because this is a larger fleet than the F100-229, deployment and basing are more complex. For details on peacetime basing, see the appendix. Table 3.3 summarizes the main points of the scenario.

The F-15s deployed on day 4 consisted of 12 each to two F-15 bases and 36 to a third base. They flew a ten-day surge on deployment (1.6 sorties per aircraft per day with a three-hour duration), followed by a 90-day sustainment period (1.0 sortie per aircraft per day with a three-hour duration).

The F-16s deployed in four waves—24 each on days 4, 8, 12, and 16 to four separate bases. As with the F-15s, they flew a ten-day surge and a 90-day sustainment schedule (from each unit's day of deployment). The F-16 sortie durations were 2.2 hours in both phases, but at sortie rates of 2.0 and 1.0 for surge and sustainment, respectively.

In all cases, we repeated the procedure we followed for the F100-229 by sizing each alternative to ensure that no required sorties were missed.[17] In each case we deployed the unit's WRE spares with the units to the FOL, where they merged to form a spares pool for all

Table 3.3

Operational Data for Aircraft with F100-220 Engines

	Aircraft		Flying Schedule			Engine Removal (per 1000 flying hours)
MDS	Total	Deployed	UTE	Surge	Sustain	(peace/war)
F-15	149	60	18	1.6	1.0	5.0/5.0
F-16	478	96	19	2.0	1.0	7.5/5.0

---

[16]There is some discussion as to whether the current higher removal rate for the F-16 will persist. As we note in more detail in Chapter Four, if we use the higher removal rates, there are not enough spares to support an MTW scenario for all but one of the repair alternatives we study.

[17]We set the tolerance level at about 2 percent (i.e., we declare no sortie loss if the overall average sortie loss is within the tolerance).

aircraft at that FOL. Figures 3.10 and 3.11 show the spares levels as a function of time for each of the four different alternatives for the F-15 and F-16, respectively. Because we now have several bases for each MDS, the graphs show the total spares in theater for each MDS as a function of days into the conflict, with one line for each alternative. Because the deployments are not simultaneous, the threshold curves are more complex and have been omitted.

As with the F100-229, the FSL outperforms the other three alternatives in both cases, with the deployed JEIM (DecDep) performing worst. As with the F100-229, this is due to the time it takes to move the JEIM into theater and get it fully operational. Note that in both cases the JEIM does provide a better recovery near the end of the war, but this is because it has more resources in rail teams, as shown in Table 3.4.

The transportation requirements for deploying an F100-220 JEIM can be computed as with the F100-229. On the basis of the Manning and

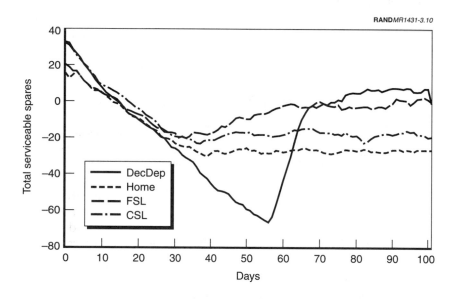

Figure 3.10—F100-220 F-15 Spares Performance

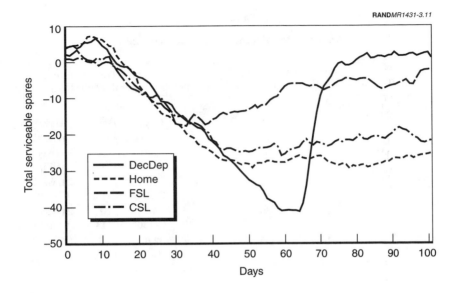

Figure 3.11—F100-220 F-16 Spares Performance

Equipment Force Packaging (MEFPAK) document, we take the F-15 JEIM to be 50 ST and the F-16 JEIM to be 25 ST for a total of 250 ST, which is 3.8 C-5 equivalents.

An analysis of the average transportation requirements for the centralized alternatives, which necessitate engine movement, can be done using average removal rates and the required flying program. The results are also given in Table 3.4. Inasmuch as the F100-220-equipped aircraft flow into theater over a period of several days and commence their surge flying at the time of entry, the transportation requirements are summarized in terms of transport aircraft equivalents for the first 1 to 14 days, the following 15 to 30 days, and then to the end of the war (when all aircraft are flying at the sustain rate).

The requirements during the first month of the war for intertheater lift is about 50 percent more than that required to move the JEIMs themselves. The intratheater lift for the FSL options is also substantial but is buying increased performance. For this engine, all but the

Table 3.4

Wartime Requirements for the F100-220

Peace	War	Rail Teams	Test Cells	Engine Transportation (aircraft equivalent)		
Decentralize	Deploy	60	34	3.8 (C-5)		
				Days 1–14	Days 15–30	Days 31–100
Decentralize	No deploy	61	22	2.8 (C-5)	4.1 (C-5)	1.6/week (C-5)
Decentralize	FSL	52	26	43 (C-130)	62 (C-130)	24/week (C-130)
CSL	FSL	30	8	43 (C-130)	62 (C-130)	24/week (C-130)
CSL	CSL	35	8	2.8 (C-5)	4.1 (C-5)	1.6/week (C-5)

FSL alternatives are unacceptable, primarily because of their inadequate support of the F-15 FOLs.

ANALYSIS OF THE TF-34

The third engine for which we analyzed repair alternatives was the General Electric TF-34, the power plant for the A-10. Unlike the F100 series, the TF-34 is a nonmodular engine and is much older than the F100 engines, having entered service in the late 1970s. In the late 1980s, the A-10 fleet was declining sharply as attention turned to the use of F-16s and F-15Es for ground attack roles. However, the performance of the A-10 in Operation Desert Storm, when it was used to great effect against Iraqi armor, prompted a reevaluation of its role, and the aircraft has thus been retained in greater numbers than had originally been planned. The A-10 was also effective in Kosovo both in a ground attack role and as a controller for other aircraft. The A-10 engine does not have an afterburner, which contributes to its higher time between removals, but because it is old, there are arguments for upgrading the engine if the aircraft is retained in the fleet.

One interesting feature of the repair concepts for the TF-34 is that it is currently mixed between centralized and decentralized alternatives. Shaw Air Force Base in South Carolina currently does JEIM repair for Pope Air Force Base in North Carolina and Spangdahlem Air Force Base, Germany, even though Shaw itself no longer has A-10s assigned. This consolidation was done because Shaw had both equipment and space when the A-10s were withdrawn, and there was a reluctance to invest heavily in extensive infrastructure reorganization given the small number of A-10s and the plans for removing the aircraft from the inventory.

For repair times, we use the empirical distribution of repair times from the Shaw JEIM.[18] Unlike the analyses for the other engines, we ran these simulations for five years (with the war starting on day 730) because the total removal rate for the TF-34 is so low (1.3 per 1000 flying hours for both scheduled and unscheduled in total) and because the in-work time for the TF-34 is substantially longer than for the F100 engines studied. This extra run length allows the simulation to reach steady state in some cases where the capacity is generous.

For the wartime analysis, we use the same single-MTW scenario as that used for the previous two engines. The A-10s deploy in units of 42, 12, 12, 24, and 15 on days 6, 11, 16, 17, and 19, respectively, to five FOLs. In all cases, only one or two bases contribute forces to an FOL, and each FOL contains either active or Guard/reserve aircraft only. As with the F-15s and F-16s, each unit flies a ten-day surge (about 1.6 sorties per aircraft per day, duration three hours), followed by 90 days of sustained flying (about 1.0 sortie per aircraft per day, duration three hours). Table 3.5 summarizes the operational data for this engine.

Figures 3.12 and 3.13 show the spares performance for the TF-34 during the MTW scenario for the active and Guard/reserve FOLs.

---

[18]Unlike the other engines, the TF-34 repair load is a mixture of "quick-turn" and more comprehensive repair. However, our analysis of CEMS and CAMS data, as well as our interviews with Shaw personnel, did not allow us to accurately estimate the proportion of engines for which each type of repair was done. We use the ENMCS distribution from the Shaw centralized JEIM because a large number of engines were repaired, and the distributions across all JEIMs of repair times and ENMCS were similar.

Table 3.5

Operational Data for Aircraft with TF-34 Engines

	Aircraft		Flying Schedule			Engine Removal
MDS	Total	Deployed	UTE	Surge	Sustain	(per 1000 flying hours)
A-10	325	105	19	1.6	1.0	1.3

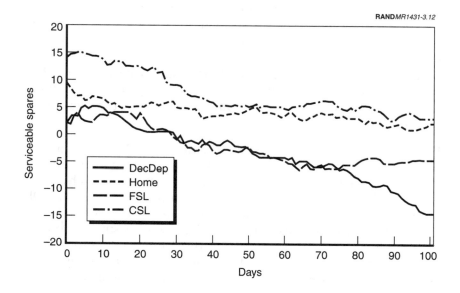

Figure 3.12—TF-34 ACC Active Spares Performance

The pattern here is much different from that for the F100 engines: For the active bases, the home support and CSL alternatives actually dominate, whereas for the Guard and reserve bases, the alternatives all give roughly equivalent performance. Further, in both cases the overall pattern is the same: a slowly declining number of spares over the course of the conflict instead of the decline/catch-up pattern seen before. This pattern is due primarily to an interaction between

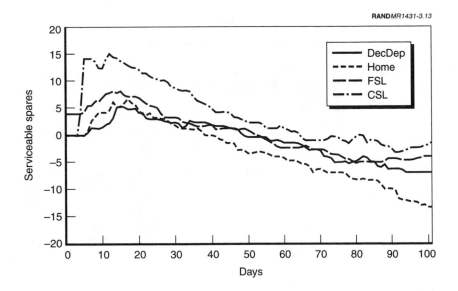

RAND*MR1431-3.13*

Figure 3.13—TF-34 Guard/Reserve Spares Performance

the number of available spares and the relatively long repair time needed to turn the TF-34. For this engine, the repair capacity is gradually soaked up with in-work engines as the war proceeds. Unlike the F100 engines, more repair capacity would help here. However, the overall performance is still good by virtue of the relatively low removal rate of this engine.

The differences between the active and Guard/reserve base patterns are due to the smaller number of WRE spares available to the latter units and, for the home case, to the difference in JEIM capacity for the active and guard bases (the latter are small, so the fragmented home capacity is not as effective). A summary of resource needs is given in Table 3.6.

The transportation requirements for the TF-34 can be computed in exactly the same way as for the F100-220. For the decentralized case, the MEFPAK indicates a size for a TF-34 JEIM of 31.2 ST. Deploying six of these would require about three C-5 equivalents. For the

Table 3.6

Wartime Requirements for the TF-34

Peace	War	Rail Teams	Test Cells	Engine Transportation (aircraft equivalent)		
Decentralize	Deploy	25	13	3 (C-5)		
				Days 1–14	Days 15–30	Days 31–100
Decentralize	No deploy	22	16	0.2 (C-5)	0.7 (C-5)	0.25/week (C-5)
Decentralize	FSL	23	18	2.5 (C-130)	8 (C-130)	3/week (C-130)
CSL	FSL	22	9	2.5 (C-130)	8 (C-130)	3/week (C-130)
CSL	CSL	20	6	0.2 (C-5)	0.7 (C-5)	0.25/week (C-5)

various centralized alternatives, the wartime transportation requirements are also summarized in Table 3.6. In this case, all of the alternatives perform acceptably.

The movements during the first month of the war are substantially less than those for the JEIM. In fact, the total movement requirement for engines is only slightly higher than the JEIM movement requirement and is concentrated after the first month.

Chapter Four

ROBUSTNESS OF JEIM ALTERNATIVES

In Chapter Three, we began the evaluation of the repair alternatives by stating the values of a substantial number of variables, such as transportation time, spares levels, removal rates, and surge period length. In most cases, we picked values for these variables that correspond to current experience because our primary goal was to evaluate how the alternatives would perform in today's world.

However, these values are not fixed. Different scenarios may arise that require different flying profiles. Moreover, removal rates may change (such rates have typically increased over time, but in Kosovo F-15Es with F100-229 engines experienced dramatically *reduced* unscheduled removal rates[1]). In this chapter we explore the effects of changes in some of these variables on the performance of the alternative repair concepts, and we then test if the relative performances we saw in Chapter Three persist. We also include a brief analysis of a centralized spares policy.

TRANSPORTATION TIME

Transportation is probably the most critical and the most contentious of all the parameters in the analyses in Chapter Three. Of all the issues surrounding centralization of repair of any kind, the requirement for new transportation and the dependability and speed

---

[1]See "USAFE P&W Operation Allied Force," HQ USAFE/LGM, undated briefing. This briefing gives statistics on removal rates and repair performance for all Pratt & Whitney engines used on U.S. Air Force fighter aircraft employed in the Kosovo operations.

of those arrangements arouse the strongest debate. This is particularly true as the Air Force spends more time in joint operations with jointly operated support and as the defense agencies continue to expand the scope of fast transportation contracts with express carriers such as Federal Express, United Parcel Service, and Emery. We must therefore evaluate how sensitive the wartime results given in Chapter Three are to varying transportation times.

Because its structure is simpler, the F100-229 is the first case we consider. In this case, aircraft are deployed to two single-MDS bases (FOLs), one with F-15s only and the other with F-16s. The two major alternatives that require transportation are the FSL (intratheater lift) and home alternatives (intertheater lift).[2]

Figure 4.1 shows the effect in terms of sorties missed of varying the one-way intratheater transportation time for scenarios with an FSL

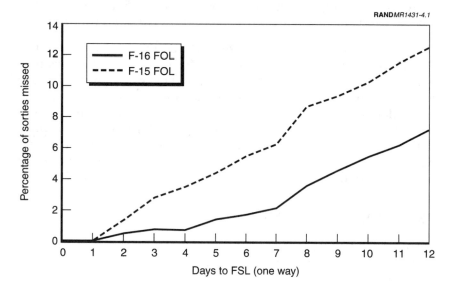

Figure 4.1—F100-229 FSL Transportation Sensitivity

---

[2]Home alternatives include both home base support (decentralized support) and CSL (centralized support) in CONUS. Since the CSL option has the same transportation time as home support, we will present only the home support results in this chapter.

supporting the F100-229. It is evident that the performance of the FSL is highly sensitive to transportation time (recall from Chapter Three that the degradation in missed sorties indicates that spares levels will be significantly worse than before). We can see that at about two days, the sorties missed are within the tolerance. However, times beyond two days for F-15s and times beyond four days for F-16s cause too many missed sorties.

Figure 4.2 shows the result of varying the one-way transportation time to the home base (intertheater lift) for F100-229 repair. Again, the performance is critically dependent on transportation and begins to degrade when the time exceeds the 15 days assumed in the analysis.

Figures 4.3 and 4.4 show the sensitivity analysis for the F100-220 engine for the intratheater (FSL) option and intertheater (home) option, respectively (note the expanded scale on the y-axis).

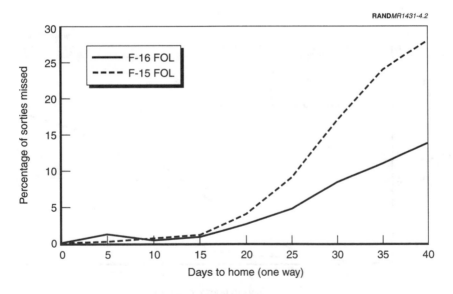

Figure 4.2—F100-229 Home Transportation Sensitivity

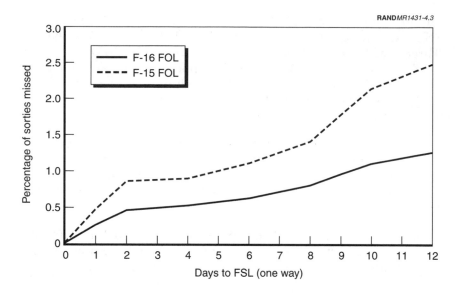

Figure 4.3—F100-220 FSL Transportation Sensitivity

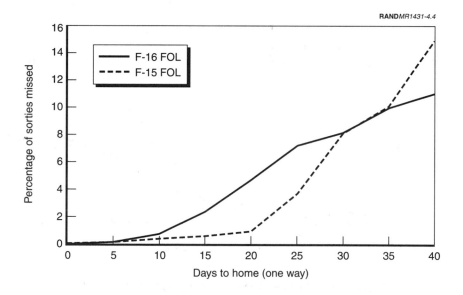

Figure 4.4—F100-220 Home Transportation Sensitivity

For the F100-220, the results are not quite as sensitive to transportation time, especially for the FSL alternative. These are averages over a number of bases, however, with somewhat different sets of spares (recall that the F-16 bases are a mix of active and Guard/reserve deployments).

As we noted in the analysis of alternatives for the TF-34, an FSL and a CSL do equally well in supporting the engine for the duration of our illustrative MTW scenario owing to the engine's reliability and the currently available spares. Figure 4.5 shows the effect of varying transportation time on the more stringent criterion of average serviceable spares[3] for the TF-34. Note that in this case we stretched the

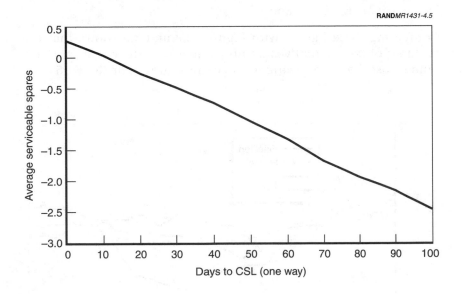

RAND*MR1431-4.5*

Figure 4.5—TF-34 CSL Transportation Sensitivity

---

[3]Average serviceable spares can be defined as the average number of serviceable spares available each day over the entire period of the conflict. It is more stringent because, as noted above, required sorties can be met even when some aircraft are grounded because there are no spare engines.

transportation time out to 100 days, the length of the MTW. Effectively, after 35 days no engines are returned during the MTW— yet the decline in average engine spares is minimal. For the TF-34, at least, transportation time does not appear to be critical (at current spares levels and removal rates).

SPARES

In Chapter Three we noted in our analysis of F100-229 repair that if only WRE spares were deployed, the serviceable spares levels dipped very close to the threshold for missing sorties. We ran a set of analyses in which we deployed *all* available spares with the engaged units. The spares performance curves for the different repair alternatives are shown in Figures 4.6 and 4.7.

Comparing these figures with Figures 3.8 and 3.9 shows that the rankings of the alternatives do not change, but the curves are now much closer together owing to the extra spares. For the F-15 FOL,

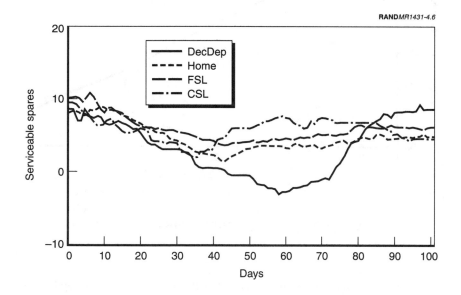

RAND*MR1431-4.6*

Figure 4.6—F100-229 F-16 Spares Performance: All Available Spares Deployed

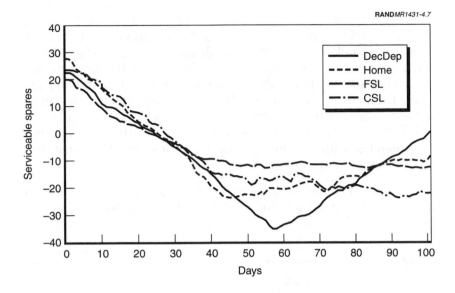

RAND*MR1431-4.7*

**Figure 4.7—F100-229 F-15 Spares Performance: All Available
Spares Deployed**

the FSL is still clearly better, but the F-16 unit does well with any alternative other than the deployed JEIM. These results do suggest
that the available spares levels are somewhat low. The F100-220 results are qualitatively similar, while the TF-34 case has acceptable
performance.

REMOVAL RATE

Another key parameter is removal rate. In the analyses in Chapter
Three, we used the removal rates that were being observed for each
engine in FY 2000. However, these rates are not necessarily constant.
The rates may go up as engines age, and there may also be periodic
sharp increases in removal rate if an urgent TCTO affects a large proportion of the fleet. Alternatively, new maintenance practices such
as reliability-centered maintenance (RCM) may substantially reduce
removal rates. For all of these reasons, it is of interest to examine the
effects of removal rate changes on the performance of different
alternatives.

As noted in Chapter Three, the removal rate used for the F100-229 was 5.0 per 1000 engine hours for both F-15s and F-16s. Figure 4.8 shows the overall loss in sorties for both aircraft in our MTW scenario as the removal rate is increased from five up to 100 (an extreme number). Each line in the figure shows the effect of such a change for the three major alternatives. The FSL alternative does very well on this robustness test in comparison to the other two alternatives. In this case, we have given each of the alternatives the same number of rail teams (the optimal number for the deployed JEIM) so that they can be more accurately compared.[4]

We conducted a somewhat more comprehensive exploration of changes in removal rates for the F100-220. The performance in

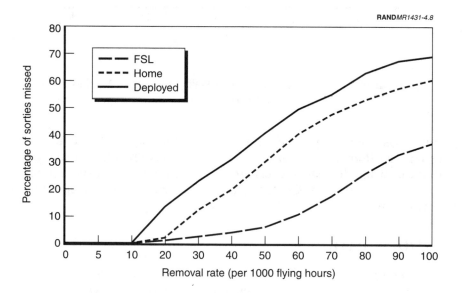

Figure 4.8—F100-229 Effect of Removal Rate Changes

---

[4]Although the FSL option continued to be the better option with its original number of rail teams, we felt that the analysis would be more sound if we varied the removal rate only while keeping all other resources fixed.

terms of sorties missed was very similar to the F100-229 results, with the FSL option performing better than other scenarios. However, a small removal-rate increase to 7.5 per 1000 engine hours led to a loss of sorties for all alternatives except the FSL option. We therefore performed a further analysis on the F100-220 system by fixing the removal rate at 7.5 per 1000 flying hours and sizing the JEIM for each alternative so that there was no queue length (i.e., they have the maximum capacity required for the combination of spares level and transportation performance). Figure 4.9 shows the comparison between the four alternatives (for F-16 FOLs) for this increased removal rate. In this case, the performance of the FSL option is only slightly worse than that shown in Figure 3.11, while the other alternatives show substantially poorer performance and are missing sorties (i.e., number of serviceable spares below the minimum threshold), with both the CSL and decentralized-deployed cases missing over 10 percent.

In order to examine the potential effects of such initiatives as RCM, which has shown some promise for substantially reducing removal rates, we reduced the removal rates for the F100-220 engines as well. The results are displayed in Figure 4.10 for the F100-220 installed on the F-15.

The "baseline" curves are for the assumed removal rate of 5.0 per 1000 engine hours. As the removal rate is successively improved, the overall performance of the deployed JEIM improves—but even at a removal rate of 2.0 per 1000 engine hours, the spares performance is worse than that of the FSL from days 50 to 65. The reduced removal rate obviously does produce better performance during the earlier part of the war (possibly the most critical time). The results for the F-16 deployed JEIM and the home case for both aircraft show qualitatively similar results in that only at the best removal rate here does the decentralized-deployed option become competitive with the FSL at the baseline removal rate.

The TF-34 removal rate is already so low that results are not sensitive to removal rates that are varied within any reasonable range.

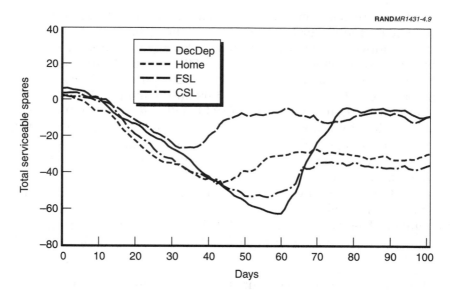

Figure 4.9—F100-220 F-16 Spares Performance Comparison with 7.5 Removals per 1000 Engine Hours

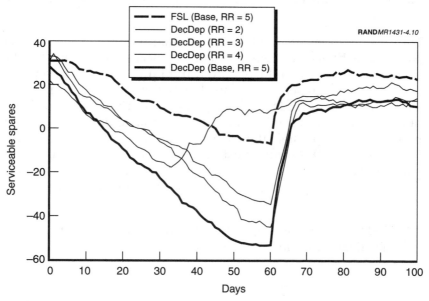

Figure 4.10—F100-220 F-15 Spares Performance of Deployed JEIM for Varying Removal Rates

LENGTH OF SURGE

Another measure of robustness that is closely related to removal rate is the uncertainty of the scenario. In this analysis, the illustrative MTW scenario is modeled closely on that used in the *Defense Planning Guidance*. However, the details of actual conflicts are uncertain: They may be more intense than expected or last longer. We would like to evaluate how well the different alternatives handle changes in the scenario.

Figures 4.11 and 4.12 show the effect of lengthening the surge period on missed sorties for the F100-229 (F-15s and F-16s, respectively). In both figures, the x-axis contains the length of the surge period, starting at ten days for the base scenario and continuing up to 50 days (one-half the length of the entire MTW). Because each alternative has been sized for a ten-day surge, it is not surprising that lengthening the surge results in more missed sorties. For both the F-15 and the F-16, however, the FSL alternative does a better job in adapting to the extended surge in that missed sorties are uniformly less for that option.

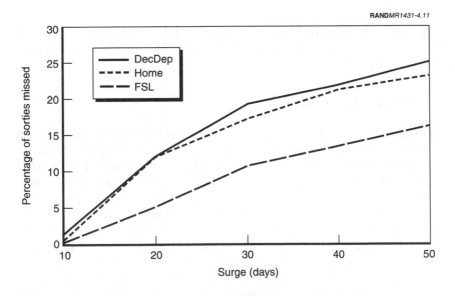

Figure 4.11—F100-229 Effect of Surge Extension for the F-15

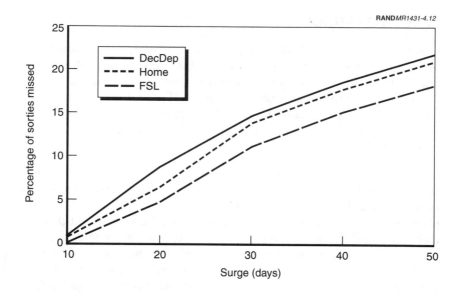

Figure 4.12—F100-229 Effect of Surge Extension for the F-16

The F100-220 results are qualitatively similar; as before, the TF-34 removal rate is so low that surge extension has no effect on the performance of any alternative.

CENTRALIZED SPARES

The final excursion addresses the issue of centralized spares levels for FSLs. It may be recalled that in the analyses in Chapter Three, we stated that the spares levels at each FOL were composed of the collection of WRE spares from each unit deployed to that location. It is possible, however, that a centralized set of spares at the FSL might well be more efficient, since any unit would get a spare shipped immediately rather than depending on its own spares while waiting for repair. Figure 4.13 shows the result of centralizing all F100-220 spares at the FSL and shipping spares to FOLs from that stockpile when a broken engine is shipped to the FSL. The "distributed" curve is the sum of the serviceable-spares curves for the FSL option from Figures 3.10 and 3.11.

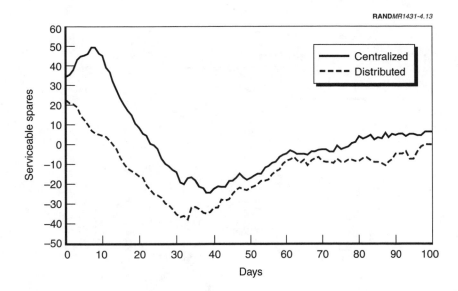

RANDMR1431-4.13

Figure 4.13—F100-220 FSL Centralized Spares

The "centralized" curve in Figure 4.13 gives the serviceable spares in theater as a function of time for the case of centralized spares stockpiles. It can be seen that the centralized curve gives substantially better performance in the early part of the conflict, with a narrowing gap during the middle of the war, when the number of available spares is simply not enough to cover the requirements.

Chapter Five

CONCLUSIONS AND RECOMMENDATIONS

From the analyses described in Chapters Three and Four (supplemented by those in the appendix for peacetime operations), we can draw a number of conclusions and recommendations. The conclusions can be divided into those drawn from the wartime scenarios and those from peacetime.

WARTIME SUPPORT

For an MTW, Deploying the JEIM to the FOL Is Slow

For each of the engines, the deployed JEIM alternative had the worst performance of all of the alternatives during the first part of the war. In each case, the number of holes was greater than for any other scenario—a deficit that would require the purchase of more spares to avoid. The problem is that the planned time to having a fully functional JEIM is too long. Further, some of that time is irreducible with current technology if the test cell is built from scratch: The concrete pad for the test cell takes 30 days to set in order to hold the thrust of a modern fighter engine. Constructing pads at many potential FOLs could reduce the time, but ultimately this decreases the flexibility of deployment required by the expeditionary operational concept.[1] Further, the deployed JEIM option takes more people—people who

---

[1] One retired engine maintainer related that he was part of a team that had constructed an "instant" test cell by burying a bulldozer and attaching steel cables to tie down the test-cell engine stand. Although extra bulldozers might not always be available, this story does suggest that there may be other available options.

have to be deployed to the FOL itself—thereby increasing the fraction of the logistics footprint moving into the area of the theater generating combat sorties.

FSL Is the Best Option for the F100-220 and F100-229[2]

The FSL option clearly dominated the others for these two engines. Owing to their relatively high removal rates, the speed with which the FSL can be brought into play and its short transportation pipeline allow it to work effectively in the MTW. An additional advantage is that it requires fewer personnel to provide better performance at the beginning of a conflict (probably the more critical time), and it can do better than the deployed JEIM at the end of a conflict by adding more capacity (but still not in excess of the deployed JEIM requirement). Further, the FSL option is more robust when faced with some uncertainties, such as changes in demand rates and extensions of the surge phase.

However, this alternative *requires* dedicated intratheater transportation. Our results show that the performance of the FSL alternative is quite sensitive to transportation times to the FSL (with current spares levels). Further, the amount of lift needed is fairly substantial, especially for the larger F100-220 fleet. Finally, the results indicate that the lift would probably be required daily, at least near the beginning of the conflict.

Either an FSL or a CSL Works for the TF-34

Because of the low removal rate for the TF-34, either an FSL or a CSL will provide adequate support. In addition, the performance of the centralized alternatives is not very sensitive to transportation times (with current spares levels).

---

[2]On the basis of our current research, we believe that two or three judiciously placed FSLs could provide service to most of the areas in which the United States has vital interests. Further, these facilities could provide ongoing support to currently deployed forces. The locations and capabilities of FSLs are a topic of current RAND research (LaTourrette et al., forthcoming).

Other Wartime Issues

Another result derived from our simulations is that current WRE spares levels may not in fact be adequate for more intense MTW operations. We have not attempted to compare our results numerically with the current engine spares process,[3] but the deep dip in spares levels during the early part of a conflict is troubling. For the F100-229, the analysis included the deployment of all spares to the FOLs, with a substantial decrease in the number of holes for all alternatives, but there were still holes for the F-15 even for the FSL option.

Although the results raise questions about the supportability of a single MTW, they further cast doubt on the ability to support two roughly continuous MTWs.

PEACETIME SUPPORT

For the TF-34, Centralized Maintenance Makes Sense

The simulation results indicate that centralizing TF-34 repair should not cause problems in meeting sorties or in keeping spare engine stocks at safe levels even when the fleet is supported in an MTW by a CSL. Further, for current removal rates and spares levels, the performance is highly insensitive to transportation times—unlike the two newer fighter engines studied here.[4] This result is directly attributable to the low removal rate of the TF-34 in comparison to that experienced by the F100-220 and F100–229—although, as noted above, this is counterbalanced to some extent by the longer time required to repair the engine and by longer ENMCS times.

This analysis therefore supports the decision to centralize maintenance at Shaw Air Force Base for some TF-34s. It further suggests that centralizing all TF-34 maintenance would not degrade

---

[3]This is computed by the Air Force Materiel Command's Propulsion Product Group at Oklahoma City Air Logistics Center.

[4]Interestingly, a TF-34 Centralized Intermediate Repair Facility (CIRF), essentially an FSL, was established at Spangdahlem Air Force Base during the Kosovo operation in part because it was feared that the transportation times back to the CSL at Shaw would be too long.

performance and would help with problems of declining skills and ENMCS. While the CSL could support one MTW and even two (if enough capacity is provided), it may be prudent to include some TF-34 repair capability in FSLs to hedge against difficulties with intertheater transportation.

For the F100-220 and F100-229, Some Centralization Is Useful

Unlike the case of the TF-34, centralization of work in peacetime for the F100-220 and F100-229 is not as clear. For the F100-229, the difference in direct resources (rail teams) is not very great. This is due to the small fleet and to the small number of bases. In addition, the higher utilization of these resources in the centralized case means that this configuration does not have much excess capacity.

For the F100-220, the differences are much more substantial, as can be seen in Table A.2. In this case, these differences are driven by the large number of small bases using this engine, since each base needs a JEIM with at least one rail team and a test cell. Given the lesser number of flying hours, both because of the smaller units and because many of these are ANG/reserve bases, one shift in the JEIM may be adequate (0.5 rail team), which would reduce the discrepancy. This does suggest that centralization may be useful for supporting small bases; such a facility could be located at one of the large bases both to reap a reduction in resources yielded by centralization and to avoid a large part of the transportation cost attributable to moving many engines from the large active-component bases.

QUALITATIVE FACTORS

The conclusions listed above are based on our modeling analysis. However, a number of qualitative factors must be considered in the decision to centralize or not centralize JEIM shops for the engines and conditions we have considered.

Given that FSLs seem to be clearly superior to other alternatives for supporting the F100-220 and F100–229 in our MTW scenario, the following issues are relevant:

- Our analysis has assumed that removal rates are the same as those currently observed. Although the FSL has some properties of robustness to inaccuracies in this assumption, an increase in removal rate would degrade all of our performance measures under any alternative. This means that the quality of flight-line diagnosis cannot be compromised if centralization removes the JEIM from a given base (whether home base or FOL). This in turn means that flight-line engine personnel must be sufficiently experienced to do as well with diagnosis as is currently the case, which may mean that the experience level on the flight line needs more attention. In particular, if the flight line requires a larger cadre of experienced engine troops when not collocated with a JEIM, some of the reductions in personnel suggested by the results in Chapter Three (and implicitly in the appendix) may be misleading.

- The control of an FSL and the responsiveness of that FSL to remote FOLs now constitute a key concern. Here the record of centralization has been mixed. Although centralized facilities such as the current Misawa F110 Centralized Intermediate Repair Facility (CIRF) and some of the ANG centralized operations have received high marks in their support of other bases, there have also been experiences with support that base commanders judged inadequate, at least initially—such as the F100 facility at San Antonio, Texas, in the early 1990s. Other questions of organizational control (e.g., major command [MAJCOM] vs. Air Force Materiel Command, active vs. ANG/reserve) are also at issue. For centralization in peacetime and especially in wartime, these issues need to be settled early and clearly.

- Given that FSLs are implemented for wartime support, if peacetime support is not centralized, the Air Force will have to ensure that the transition from one structure to another is smooth. This will require the rethinking of areas such as information systems, command relationships, and communication requirements and, most important, practicing real centralized maintenance on appropriate occasions, such as exercises and deployments. A different system in peace and war would go against the adage to "train as you fight," and careful attention would have to be paid to ensuring that the transition was smooth.

- We note that there are several advantages claimed for centralization that we have not addressed in our modeling and analysis. For example, centralization is thought to help both training and ENMCS. We have no quantitative data on the size of these effects, and their absence here means that our analysis is conservative: To the extent that centralization can yield gains in these areas, the advantages for centralization are larger than those we have presented.[5]

RECOMMENDATIONS

Given the analysis presented in this report and the conclusions drawn above, we make the following recommendations:

1. Develop engine FSLs for wartime support of fighter engines with removal rates in the ranges experienced by the F100-220 and F100–229 (i.e., roughly all current fighter engines that have afterburners). The analysis shows that they are required to support MTWs; once in place, they can support smaller expeditionary operations with the added benefit of reducing the deployed footprint for such operations. We emphasize, however, that the development of such installations requires considerable forethought and preparation as well as planning for support from a global, strategic perspective.[6]

2. If the first recommendation is implemented, the analysis shows that it is essential that a responsive transportation system be provided both to move serviceable engines from the FSL to the FOL and, of equal importance, to return unserviceable engines to the FSL to be repaired.

---

[5]We do have one piece of anecdotal evidence for a training advantage of FSLs: A senior noncommissioned officer (NCO) at one of the JEIM shops we visited pointed out that because there was some limited centralization in his MAJCOM, deploying units that helped staff the engine CIRF could send some relatively junior people because the CIRF had enough depth to supervise them. This both eased the deployment burden on senior people and allowed more senior people to remain at home station to supervise the junior people who were left behind.

[6]See Tripp et al. (1999).

3. The analysis (as well as empirical evidence) supports the centralization of TF-34 repair even to the extent of using CSLs to support MTWs. As a hedge against uncertainties with intertheater transport, some TF-34 repair capability might be included in an FSL. This has no effects on performance and would help conserve the declining skill base for the TF-34 as the A-10 is phased out of the inventory.

4. The centralization of peacetime repair for small bases with F100-220s and F100–229s should be considered. This is where most of the gain due to centralization could be achieved. If a centralized facility is located close to the larger bases that use these engines, substantial fractions of the transportation costs estimated here can be avoided.

PEACETIME ANALYSIS

OVERVIEW

The main body of this report has focused on the performance of centralized and decentralized alternative repair concepts in a notional MTW scenario. In this appendix, we study the differences between centralized and decentralized repair for peacetime flying.

Evaluation Design

As with the MTW analysis, we proceed by using the model to establish the repair capacity required so that each alternative we consider provides good performance on our key metrics. We then rank the alternatives by capacity. One performance metric is again the requirement that no peacetime sorties be missed because of lack of serviceable engines. Because the peacetime flying program does not vary dynamically, the dynamic spares level is not of interest, so we require that each alternative also be able to maintain a positive average spares level.

The rest of the parameters, as with the wartime analysis, are set to their current levels:

- Engine removal rate.

- Repair time (based on peacetime work schedules).

- Total spares (no distinction is made between WRE and other spares).

- For the centralized alternative, CONUS transportation time is set to two to four days.

- Peacetime utilization rates for each aircraft MDS were as given in Chapter Three.

The sources of these parameter values are given in Chapter Three.

For the peacetime runs, we used a two-year run with only peacetime sorties required. The average sorties missed and engine spares levels are averaged over the two-year run.[1]

Repair Alternatives

In peacetime, there are only two repair alternatives to be considered:

- **Decentralized.** This is the current case, in which each base has a JEIM with adequate capability to support the forces stationed there.

- **Centralized.** In this alternative, a CSL does all JEIM work in one centralized facility. As before, we also assume that the CSL repairs engines for all MDS sharing an engine type and can supply each base equally well with the right configuration. Engines are returned to the originating base; there is no central spares level that immediately provides a base with a serviceable spare when one of its engines is shipped to the CSL.[2] Finally, the performance of the centralized CSL is identical to that of the base JEIM, both in repair time and in ENMCS performance.[3]

---

[1]We did some experiments in which the model was run for an initial "burn-in" period to reach a steady state; the data from the burn-in were eliminated from the analysis. However, these results did not differ from those using a two-year run without a burn-in.

[2]A central spares level would provide better support while also being more cost-effective in that the spare and the unserviceable engine could be transported in a single round trip of one truck. However, if a CSL is supporting active, Guard, and reserve units, the policy assumed here is reasonable based on current practice.

[3]It has been argued that a CSL would improve both repair time, because of its concentration of experience, and isolation from ancillary duties with which the propulsion flight is often tasked. Similarly, increased scope for cannibalization should improve ENMCS performance. These assumptions are therefore conservative.

ANALYSIS OF THE F100-229

As noted in Chapter Three, the fleet size for the F100-229 is rather small, and the largest current single unit is located outside of CONUS. In order to make our analysis informative, we constructed a "synthetic" peacetime basing structure:

- Two F-15 bases of 18 and 48 primary authorized aircraft (PAA), respectively.

- Two F-16 bases, each with 18 PAA (with slightly differing spares levels).

This is essentially the entire current F100-229 fleet considered as a single CONUS population.

Decentralized

The first task of analyzing the base case is to determine how many rail teams must be located at each base to support the peacetime flying schedule. Figure A.1 shows the percentage of sorties missed for different numbers of rail teams at each base. On the basis of this figure, we see that ten rail teams are required: four at the large F-15 base and two at each of the smaller bases. Each base also requires a test cell for a total of four.

As noted above, we also want to assess alternatives on how well each can sustain the spares levels. In Figure A.2, we plot the average spares levels from the two-year run for different numbers of rail teams at each base. We see that the configuration that satisfies the sortie requirements also keeps average spares levels in a positive steady state.

CONUS Support Location

For the small F100-229 fleet, one CSL site will be sufficient. As with the base-case analysis, we begin by finding the number of rail teams at the CSL that allows all bases to meet their sortie schedule. Figure A.3 shows the results.

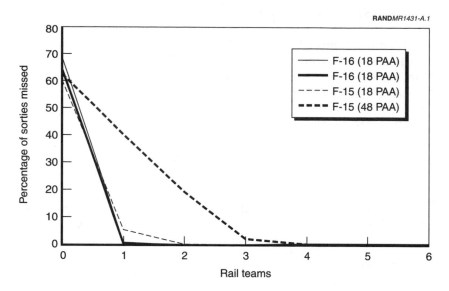

Figure A.1—F100-229 Peacetime Sorties Missed as a Function of Decentralized Repair Capacity

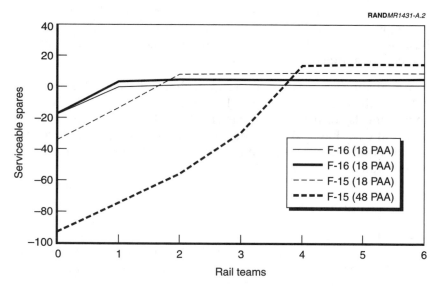

Figure A.2—F100-229 Peacetime Spares as a Function of Decentralized Repair Capacity

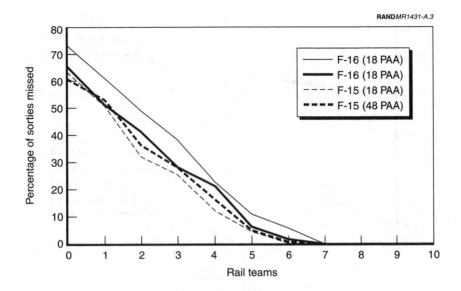

Figure A.3—F100-229 Peacetime Sorties Missed as a Function of Centralized Repair Capacity

In this case, eight rail teams are sufficient to allow all bases to meet their required sorties. One test cell at the CSL is sufficient to handle the test load for the F100-229 population. This number of rail teams is also sufficient to maintain a positive average spares level, as shown in Figure A.4.

For the CSL, however, we also need to assess transportation resource requirements. For peacetime flying activity, the average monthly number of engines going to the JEIM is 23. It is clear in the United States that sufficient transportation resources are available. The cost depends on specific origin-destination pairs, whether or not a load is guaranteed for the round trip, and the like, but a reasonably conservative estimate of a one-way engine movement in the United States is about $1200. The total is $662,000 per year in transportation costs.

Figure A.5 shows that "sorties missed" is insensitive to trip time up to six days for the larger F-15 base; this time is well within the capabilities of domestic transport between almost any two points in the United States.

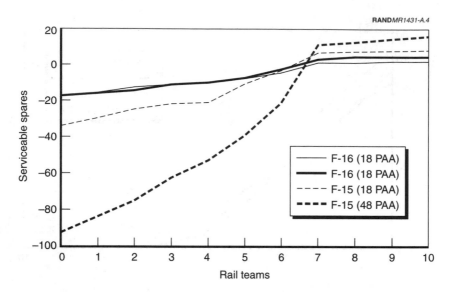

Figure A.4—F100-229 Peacetime Spares as a Function of Centralized Repair Capacity

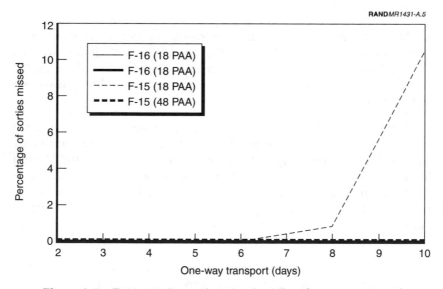

Figure A.5—F100-229 Peacetime Sorties Missed as a Function of Transportation Time to CSL

The driver of the difference in rail teams and test cells between the centralized and decentralized cases is, as might be expected, resource utilization. Although the rail teams at the large F-15 base are fairly heavily loaded (with a utilization of 91 percent), at the F-16 bases the utilization is only 36 percent. By contrast, at the CSL the rail teams average 82 percent utilization. Similarly, the four test cells supporting decentralized JEIMs have utilizations ranging from 7 percent to 33 percent, while at the CSL the utilization is 74 percent.

The peacetime analysis for the F100-229 is summarized in Table A.1.

Table A.1

Peacetime Requirements for the F100-229

Option	Rail Teams	Test Cells
Decentralized	10	4
CSL	8	1

ANALYSIS OF THE F100-220

With a larger fleet, we elected to use the current CONUS configuration of the fleet for the peacetime analysis. The basic structure is as follows:

- Three active F-15 bases with 19, 38, and 92 PAA, respectively.

- One active F-16 base with 42 PAA.

- One active Air Education and Training Command (AETC) F-16 base with 166 PAA.

- Two Air Force Reserve (AFR) bases with 17 PAA.

- 13 ANG F-16 bases with 17 to 24 PAA.

As with the F100-229, we proceed by estimating the resources (rail teams) needed to keep all units from missing peacetime sorties and to keep positive engine spares levels in each of the different alternatives. As this process is essentially the same as for the F100-220, we omit the graphs here. Table A.2 summarizes the results.

Table A.2

Peacetime Requirements for the F100-220

Option	Rail Teams	Test Cells
Decentralized	49	22
CSL	30	4

The most immediate difference between the F100-220 results and those for the F100–229 lies in the substantial differences in rail units and test cells. This is a result of the dispersion of a substantially larger fleet among many small bases with only two to three bigger ones. Since each of the small bases needs a single rail team, this drives up the difference. It might be argued that they need only half a rail unit (one shift), which would substantially decrease the difference.

An analysis of expected engine removals shows that this engine has an average of 134 removals per month. Using the transportation costs for the F100-229, this equates to a peacetime transportation bill of $3.86 million per year for CONUS units.

ANALYSIS OF THE TF-34

For the peacetime analysis, we use the current force configuration in CONUS with the addition of the A-10s at Spangdahlem, since they are supported by Shaw Air Force Base. As with the F100-220, the configuration we examine includes ANG and AFR units. The detailed configuration is as follows:

- One active wing of 72 PAA.
- One active wing of 46 PAA.
- One active wing of 20 PAA.
- Two active wings of 16 and 11 PAA for training and test.
- Two AFR wings with 17 PAA.
- Six ANG wings with 16 PAA.

For the TF-34, we model the base case as including Shaw support of Pope and Spangdahlem.

The analysis proceeds as before, sizing the resources needed for each case. As with the F100-220, we do not present the analysis in detail but give only the results shown in Table A.3.

Table A.3

Peacetime Requirements for the TF-34

Option	Rail Teams	Test Cells
Decentralized	23	14
CSL	20	3

Note that the 14 test cells for the decentralized case do not include test cells at Pope and Spangdahlem Air Force Bases, which are not used for JEIM work.

Using the average removal rate and peacetime flying schedule, the expected number of engines entering the JEIM per year is 238, which equates to a total transportation cost of $570,000.[4]

The results for this engine are somewhat surprising. There are two factors that would indicate that the CSL should be able to function well with substantially fewer rail teams than the decentralized option:

1. The fleet is more fragmented than the 229 fleet (with several small bases, each with its own JEIM).

2. The engine is much more reliable than either the F100-220 or the F100-229.

These factors are somewhat counterbalanced, however, by the longer time the engine takes to repair. This is due both to a longer in-work time and to a longer ENMCS (attributable to the age of the engine and to a corresponding difficulty in getting parts).

---

[4]This does not reflect the costs paid for shipping engines from Spangdahlem.

BIBLIOGRAPHY

Amouzegar, Mahyar, and Lionel A. Galway, *Engine Maintenance Systems Evaluation (En Masse): A User's Guide* (forthcoming).

Berman, Morton B., Irving K. Cohen, and Stephen M. Drezner, "The Centralized Intermediate Logistics Concept: Executive Summary," Santa Monica: RAND, unpublished research, 1975.

Carrillo, Manuel C., and Raymond Pyles, *F100 Engine Capability Assessment in PACAF: Effects on F-15 Wartime Capability*, Santa Monica: RAND, N-1795-AF, 1982.

Davis, Richard G., *Immediate Reach, Immediate Power: The Air Expeditionary Force and American Power Projection in the Post-Cold War Era*, Washington, D.C.: Air Force History and Museums Program, 1998.

Feinberg, Amatzia, Hyman L. Shulman, Louis W. Miller, and Robert S. Tripp, *Supporting Expeditionary Aerospace Forces: Expanded Analysis of LANTIRN Options*, Santa Monica: RAND, MR-1225-AF, 2001.

Galway, Lionel A., Robert S. Tripp, Timothy L. Ramey, and John G. Drew, *Supporting Expeditionary Aerospace Forces: New Agile Combat Support Postures*, Santa Monica: RAND, MR-1075-AF, 2000.

Hosek, James R., and Mark Totten, *Does Perstempo Hurt Reenlistment? The Effect of Long or Hostile Perstempo on Reenlistment*, Santa Monica: RAND, MR-990-OSD, 1998.

Killingsworth, Paul, Lionel A. Galway, Eiichi Kamiya, Brian Nichiporuk, Timothy L. Ramey, Robert S. Tripp, and James C. Wendt, *Flexbasing: Achieving Global Presence For Expeditionary Aerospace Forces*, Santa Monica: RAND, MR-1113-AF, 2000.

LaTourrette, Thomas, Donald Stevens, Amatzia Feinberg, John Gibson, and Robert S. Tripp, *Supporting Expeditionary Aerospace Forces: Forward Support Location Options*, Santa Monica: RAND (forthcoming).

Peltz, Eric, Hyman L. Shulman, Robert S. Tripp, Timothy Ramey, Randy King, and John G. Drew, *Supporting Expeditionary Aerospace Forces An Analysis of F-15 Avionics Options*, Santa Monica: RAND, MR-1174-AF, 2000.

Richter, Paul, "Buildup in Gulf Costly: Expenses, Stress Surge for Military," *Los Angeles Times*, November 17, 1998.

Ryan, Michael E., *Evolving to an Expeditionary Aerospace Force*, Commander's NOTAM 98-4, Washington, D.C., July 28, 1998.

Taylor, William W., S. Craig Moore, and Charles Robert Roll, Jr., *The Air Force Pilot Shortage: A Crisis for Operational Units?* Santa Monica: RAND, MR-1204-AF, 2000.

Tripp, Robert S., Lionel A. Galway, Paul S. Killingsworth, Eric L. Peltz, Timothy L. Ramey, and John G. Drew, *Supporting Expeditionary Aerospace Forces: An Integrated Strategic Agile Combat Support Planning Framework*, Santa Monica: RAND, MR-1056-AF, 1999.

Tripp, Robert S., Lionel A. Galway, Timothy L. Ramey, and Mahyar Amouzegar, *Supporting Expeditionary Aerospace Forces: A Concept for Evolving the Agile Combat Support/Mobility System of the Future*, Santa Monica: RAND, MR-1179-AF, 2000.